Drawing Geological Structures

The Geological Field Guide Series

Drawing Geological Structures

Jörn H. Kruhl

Technische Universität München, Munich
Germany

Translated from the German by Tiana Stute

Registered Office(s)
John Wiley & Sons, Inc., 111 River Street, Hoboken, NJ 07030, USA
John Wiley & Sons Ltd, The Atrium, Southern Gate, Chichester, West Sussex, PO19 8SQ, UK

Editorial Office
9600 Garsington Road, Oxford, OX4 2DQ, UK

For details of our global editorial offices, customer services, and more information about Wiley products visit us at www.wiley.com.

Wiley also publishes its books in a variety of electronic formats and by print-on-demand. Some content that appears in standard print versions of this book may not be available in other formats.

Library of Congress Cataloging-in-Publication Data

Names: Kruhl, Jörn H. | Stute, Tiana, translator.
Title: Drawing geological structures / Jörn H. Kruhl, Technische
  Universität München, Munich, Germany ; translated from the German by
  Tiana Stute.
Description: Hoboken, NJ : John Wiley & Sons, Inc., 2017. | Series:
  Geological field guide ; 7180 | Includes bibliographical references and
  index.
Identifiers: LCCN 2017007264 (print) | LCCN 2017008437 (ebook) | ISBN
  9781405182324 (pbk) | ISBN 9781119387237 (pdf) | ISBN 9781119387244
  (epub)
Subjects: LCSH: Geology–Charts, diagrams, etc. | Drawing.
Classification: LCC QE33.2.C5 K78 2017 (print) | LCC QE33.2.C5 (ebook) | DDC
  551.8–dc23
LC record available at https://lccn.loc.gov/2017007264

Cover Design: Wiley
Cover Image: Courtesy of the author

Set in 8.5/10.5pt, TimesLTStd by SPi Global, Chennai, India.
Printed in the UK

CONTENTS

CONTENTS

ABOUT THE AUTHOR

Jörn H. Kruhl is retired professor of geology at Technische Universität München, Germany. He received his Dr. rer. nat. from Rheinische Friedrich-Wilhelms-Universität Bonn and held appointments in research and teaching at universities in Mainz, Salzburg, Berlin and Frankfurt/M. For decades he worked on rock structures in numerous regions and orogens, from macro to micro, in the field and at the microscope.

PREFACE

Drawing is not one of my strengths. My first attempts at representing rock structures from the field on paper ended in disaster, both with regard to the graphic quality as well as the information content of the drawings. It took a long time for me to be able to draw structures halfway precisely and increase their recognition value; to use perspective; to learn to draw symbolically; and to compose larger outcrop drawings and block diagrams. In my early field days, Gerhard Voll accompanied me as role model and companion who showed me, by way of his own geologically and artistically sophisticated drawings, how it's done. In later field and drawing classes, many of my students were at a higher level from the start than I was when I began. Fortunately, talent is not required for geological drawing—only the willingness to observe and a little practice to acquire the basic rules for drawing geological structures.

A special merit of drawing is that it requires us to look closely. The click of the camera cannot do this. While we are drawing, we must already geologically assess what we are drawing. Therefore, not just the drawing, but also the path to it, is relevant. Graphical representation—manual drawing—is nothing old-fashioned and superfluous, or just a nice pastime. In the digital age, it is urgently needed, because it teaches us to observe and reflect, and it leads to concentration and mindfulness. This book is about drawing as a language—a language in which geological information can be conveyed precisely and straightforwardly. Contrary to art, a geological drawing is not open for a personal point of view. It is intended to capture a structure's geological message and represent it so that it can be correctly interpreted by the observer. Furthermore, drawings can be used to effectively and quickly store the geological information contained within a structure. Conversely, drawings give us quick and easy access to information while also providing us with a highly informative archive. This book is an exercise book. Nevertheless, it contains only a few, exemplary exercises. This is because, apart from the basics, geological drawing can only be learned to a limited extent with the help of dry training. One learns by observing structures in rocks and under the microscope, and by drawing them directly. For effective practice, one must go out into the field or to the microscope, or, if necessary, search for suitable samples in the geological collection. This is the only way to observe structures in different ways—by walking around the rocks in an outcrop, looking at them up close and from far away, using a hammer and chisel to expose important surfaces, or by varying magnifications and other conditions under the microscope.

This book is intended to stimulate such practice and be a companion that exemplifies, by means of different rocks and structures, the many possibilities of drawing and its development stages from start to completion. A further focus of this book is how drawings can be optimally composed with regard to high information content and quick access to this information. In addition, this book is intended to serve as an encouragement to apply drawing in daily practical work—including areas beyond those discussed here!

This is also important: The geological sign language presented in this book is not an unalterable set of rules. Like every language, it is flexible and open to change. Although the foundation of drawing may stay the same, every person can interpret the rules in their own way, develop new schemas of drawing, and arrive at their own "dialect."

For didactic reasons, many drawings have been revised or completely redesigned for this book. Several drawings, however, were taken directly from my field books or microscopy notes—with thick lines, mistakes, and corrections. They are unclean and don't always follow the rules, but reflect the real situation when drawing is a daily work instrument. Whenever possible, these drawings are depicted in their original scale, or at least not greatly reduced. They shall not be made prettier than they are. Even though it is alright and good (and even necessary for certain purposes) to make clean and aesthetic drawings, the hasty and coarse line is the normality when drawing from nature. The quick, rudimentary sketch is the colloquial language of geological scientific work. This is also covered in this book.

The foundations of this book arose mainly through my own practical work in the field and at the microscope, but also through numerous microscopy and several drawing courses for which I could occasionally, despite crammed curricula, find the time. The interest of the students as well as the colleagues was always motivating. Thank you for that.

In addition, my thanks go to Tom Blenkinsop, who encouraged me to write this book; Herbert Voßmerbäumer, who, as reviewer and with his positive attitude, helped to get the book started; as well quite a few anonymous reviewers who spoke generously and favorably about the project. The book manuscript clearly benefited from the careful inspection by Uwe Altenberger, Annette Huth, and Matthias Nega. I deeply thank you for this. Last but not least, I would like to thank the Wiley-Blackwell staff, who have accompanied the book, with great patience, through its different stages of development over the years—especially Ian Francis, Kelvin Matthews, and Delia Sandford, who accepted, with friendly serenity, my numerous excuses for why the manuscript was still not finished, and Sanjith Udayakumar, Ramprasad Jayakumar and Arabella Talbot, who supervised the book in the production phase.

Jörn H. Kruhl
Munich, January 2017

1

INTRODUCTION

"Of course you should draw! You should draw everything that can be drawn..."

"But, Professor, I have no artistic talent!"—"You do not need it! You aren't supposed to make art, but simply draw as well as you write. Firstly, so that you can learn to better see and observe, because the drawing pencil forces the eye to look closely and give a detailed account of the facts, for drawing is guided seeing; secondly, because drawing is often the shortest and best form of description. For this you need no talent, only diligence and a little guidance..."

(Hans Cloos, 1938)

Drawing is one of the elementary human abilities. It requires practice. But one must not draw with the skill of a Leonardo da Vinci or an Albrecht Dürer to be able to create drawings that are informative, aesthetic, and a joy to others. The drawing of geological objects is at a level that anyone can reach with a little practice and by following a few rules (Figure 1.1).

When we talk about drawing, we usually mean *artistic drawing*. In the case of Leonardo da Vinci—the brilliant painter, sculptor, architect, naturalist, and engineer of the Renaissance—this includes *technical* or *scientific drawing*. But in later times, the artists were rarely scientists and the scientists rarely artists. The tasks were distributed. Alexander von Humboldt "measured the world" and Aimé Bonpland drew it. Carl von Linné systematized species classification, while Maria Sibylla Merian painted insect and flower pictures, and John James Audubon left behind "The Birds of America." Only a few artists sketched geology (apart from the omnipresent Goethe), like Robert Bateman, for example: "I enjoyed painting the rock, a kind of granite called gneiss, using little trickles of turquoise and pink and yellow and gray. When I paint rocks I like to convey their characteristics and to make sure that they belong in the landscape and are recognizable geologically" (Terry, 1981); it is the geoscientists, rather, like Clarence E. Dutton (1882) or Albert Heim (1921), that have seen rocks and their structures with the eyes of artists (Figure 1.2).

Today, constructive drawing is what is meant by the term technical drawing, and that is done almost exclusively by computers. Academic (or scholarly), in particular scientific and specifically geological, drawing resists automation, because

Drawing Geological Structures, First Edition. Jörn H. Kruhl.
© 2017 John Wiley & Sons Ltd. Published 2017 by John Wiley & Sons Ltd.

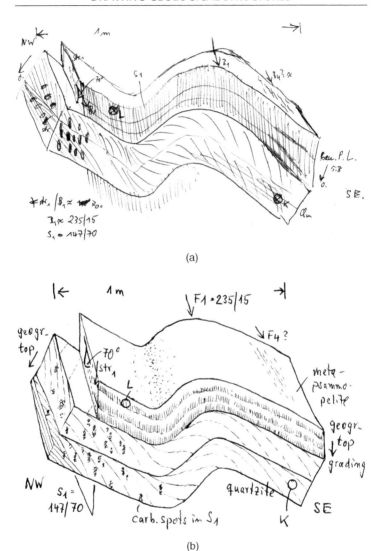

(a)

(b)

Figure 1.1 *One of the author's early, but failed, attempts to draw samples and outcrops in the field, and a better version of the same drawing.* **(a)** *Monoclinic fold in psammopelite and quartzite layers of the Moinian (Grampian Highlands at Loch Leven, Scotland); field drawing; outcrop KR513; field book 6 (Kruhl, 1973). The drawing contains numerous shortcomings; above all, imprecise layout of lines, a sloppy perspective, and an incorrect positioning of foliation planes in the metap-sammopelitic layers.* **(b)** *The same drawing redrawn years later. Cross bedding and S1 foliation planes are more precisely placed; the perspective is correct and, consequently, the 3D appearance of the drawing is better; the carbonate spots are more realistically illustrated; and the labeling is more closely related to the struc-tures. Circles L and K indicate positions of samples. Both drawings ca. A6; black ballpoint pen.*

Figure 1.2 *Drawing of part of the Grand Canyon, "Vishnu's Temple" (Dutton, 1882, plate XXXIV): a felicitous combination of artistic, geological, and geomor-phological representation.*

3

nature knows no straight lines. Geological objects, like rock layers, folds, volcanic dykes, foliation planes, joints, and the outlines of crystals cannot be represented using the shapes of Euclidean geometry. This is not a question of precision, since the shapes of all these objects don't just vary by chance from the Euclidean form. We know today that many natural processes are not linear and produce shapes of non-Euclidean, fractal geometry (Mandelbrot, 1983). Many geological shapes appear complex and are usually described qualitatively (*sutured, rounded, amoeboid*) or are represented with the help of picture plates, like the degree of rounding of sand grains, for example. These images are usually paired with specific names (*angular, subangular, subrounded, rounded*) to ensure the transition to a written description. Complex structures can be captured truly precisely only when they are quantified using fractal geometry. Using these rules while sketching geological structures is well worthwhile. The gain in naturalness and closeness to reality is big.

While scientific drawing is based on a number of rules of artistic drawing, it has many of its own laws. Therefore, geological drawing requires different rules, in part, from artistic drawing. However, the principally irregular form of geological objects does not necessarily mean that it must always be drawn "irregularly" or "fractally." There are reasons for schematic, Euclidean drawing. This is why geological drawing must shuttle between lifelike and abstract representation. This is not easy, and the questions of when is it better to draw realistically, when is an abstract representation more effective, and how can a balance be established between the two, will be discussed in detail.

What is drawn must, however, already be technically understood and interpreted. This is the only way to select and distinguish between what is geologically important and unimportant. "It is the theory that determines what we can observe" (Einstein, 1955). Or, in other words: "You only see what you know" (Weizsäcker, 1955). When transferred to the drawing of geological structures, this means: We only see what we already have as a mental model. We only see the geological structures we expect and that already belong to our knowledge base. Although this may seem a little bit strict, it is true that we have difficulty perceiving and often dismiss structures that we do not know and that aren't part of our empirical knowledge.

Of course it is fundamentally possible to perceive even the unexpected or unusual, but it's hard, and we therefore do well to look at structures exactly before drawing them. If we interpret first, it will be easier to perceive the unexpected and unusual, and incorporate it into our knowledge and experience. This can be time consuming, and causes difficulties. Nevertheless, drawing itself, the physical process of seeing and sketching geological objects, is on a level of craftsmanship that anyone can achieve with a little practice, and, in any case, a "bad" drawing is still better than no drawing!

There are some aspects of geological drawing relating to geological maps and the construction of profiles that we will not touch upon, because they veer too much into the field of technical drawing. For this, there are a sufficient number of good books and, above all, websites where these techniques can be trained online. Furthermore, this book is not about drawing fossils. Although the drawing of fossils coincides in many ways with the drawing of geological structures, there are still some fundamental differences, like the object fidelity, which is essential to the drawing of fossils but more of a hindrance when drawing geological structures. The present book is mainly about:

- the way in which one must represent geological objects at different scales,
- how the purpose of the representation affects the nature of the representation,
- the way in which a balance between detailed and symbolic representation must be maintained in such drawings, and
- how one can practice all of this.

We will go from small to large, from thin section to outcrop, especially the ensemble of outcrops, and from the two-dimensional representation to the three-dimensional. This order has been chosen in part because two-dimensional representations are technically and in their principles easier, and because the two-dimensional surfaces of three-dimensional objects are seen first. Secondly, big geological objects are made up of many small pieces, and the bigger picture is best understood, if we understand the details.

This book is intended as an exercise book for the purpose of self-study. It should encourage the playful retention of structures, the anchoring of one's own geological data collection in the form of graphic representations, and the occasional replacement of the camera with paper and pen in the field. In addition, this book is meant to encourage the use of the benefits of exact drawings especially when it comes to precision and conciseness (e.g., in publications). Finally, I would like the representations in this book to show that geological structures have not only scientific value but also deserve our attention for their complexity and aesthetics.

1.1 Why Do We Need Drawings?

Anyone who has tried to describe a thin section, a rock sample, or an outcrop without the help of drawings (or photos), would probably not pose such a question. Compared to the expressiveness and the rich detail of graphic representations, the spoken or written word is an inadequate tool. Drawings and photos can document things that would otherwise take much more time to describe, in no time. And since graphics can be digitized, the electronic storage and processing of graphic information is not a problem.

There is no strong conflict between drawing and photo. Photography is a quick and easy type of documentation. When taking pictures, one can be sure that all the details in the range of resolution are preserved. Even on a small scale, subtleties, which would not be accessible in a drawing or which would cost a lot of time to include, can be captured. Those who have participated in a field trip with an eager leader racing through a packed itinerary have learned to appreciate the camera. If 20 minutes are allotted to an outcrop with nice sedimentary structures or complex folds, one can leave the field book or the sketch pad in one's pocket, unless of course one belongs to the small, gifted group of precise fast-drawers.

On the other hand, the photograph reaches its limits when it comes to filtering out the essentials from a jumble of small details. Who hasn't photographed an outcrop that appeared clear and impressive to the eye, or even a rock thin section, that then, in black and white or in color, on paper or on the screen, appeared only as an indissoluble mixture of details distorting the essentials? In addition, the photo also captures the unimportant surrounding and provides information that we do not need and that we must carry with us as interfering ballast. Computer-aided photography provides opportunities to smooth surfaces and thus to convert photographs overloaded with details into sketch-like representations (Hayes, 2008). This development, however, is still in its early phases and it remains to be seen how useful it will be for geological objects with complex, detailed structures, and, especially, if the effort of editing the photographs exceeds that of sketching.

For three-dimensional objects, it may prove especially hindering that a photo only delivers a two-dimensional view of structures that in small form, even at the two-dimensional outcrop face, appear three-dimensionally and contain much more information in three dimensions than in two (Figure 1.3). We can

---

Figure 1.3 **(a)** *Photo of the "Spitznack" Fold (near the Loreley, Middle-Rhine region, Rhenish Massif, Germany). Centimeter- to decimeter-thick metapsammopelitic layers are bent to an open, monoclinic fold. The boundaries of bedding are represented by strong fractures in the horizontal fold limb and by weak fractures and differences in brightness in the vertical limb. In addition, a schistosity can be recognized. It is represented by narrow-spaced, nearly parallel fractures. The schistosity is pronounced and steep in the horizontal limb, fan-shaped in the fold crest, and is barely visible in the steep limb. Additional fabric details cannot be recognized. Hammer as scale.* **(b)** *Schematic drawing of the same fold. Based on the small protrusions and the fabrics that can be recognized on them, the planar, 2D view has been supplemented to form a 3D block. Highlighted are (i) the partitioning of the schistosity to two different sets of foliation planes in metapsammitic layers, (ii) the pile shape of the schistosity in metapelitic layers, (iii) the stretching lineation on an obviously bedding-parallel foliation plane, (iv) the compressed and sheared quartz veins in the steep limb, and (v) the slickensides on steep bedding-parallel shear planes. All these structures are not visible in the photo and are only revealed by close observation of the fold. Modified after Zurru and Kruhl (2000, Figure 33); size of original drawing ca. B4. A more comprehensive drawing is shown in Figure 4.27.*

(a)

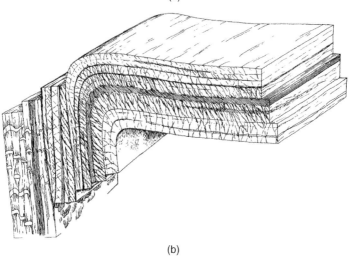

(b)

(interpretatively!) capture three-dimensional structures more clearly and comprehensibly on two-dimensional paper when drawing perspectively. This also applies during microscopy: The brain can better distinguish between important and unimportant, and filter out the structures in question. How aesthetically and clearly identifiable the deformation and metamorphic fabrics could be, were it not for the retrograde influences that transform aesthetic plagioclase twins into an "ugly" mix of small mineral grains or that superimpose all the striking quartz deformation structures with an iron-plated net of micro fractures.

This means: Drawings are important when it comes down to emphasizing details or omitting the unessential, and when representing the three-dimensional three-dimensionally. This is a scientific decision! The strengths of the sketch are highlighting, omitting, and combining things that were not together in the original. One composes an image so that it meets the necessary requirements. This may seen "unscientific" at first glance. But fidelity is not most important; what is worth striving for is the scientific documentation and the interpretation of structures. Of course, structures that do not belong together should not be put together in such visual compositions, and no other, or even false, interpretations may emerge. But a drawing is not purely for documentation; it communicates beliefs and ideas. From this follows: Drawings must always contain an interpretation. Strictly speaking, one can hardly draw without interpreting, because even an omission is an interpretation. The geological stereogram (Chapter 5) is one of the most impressive forms of interpretation. In it, detailed pictures from individual outcrops are combined into one image that is neither to scale nor must show things next to one another as they occur in nature. The point of a stereogram is to represent the principle of the large-scale structure of geological area. The stereogram is therefore the model that a processor makes of a given area.

Drawings assist geologists in their everyday work. Sketches preserve the memory of details that would've been impossible or too complicated to photograph in the first place (poor lighting conditions, too many details, vegetative coverage, etc.). Drawings become indispensable when it comes to defending one's hypotheses in discussions and when trying to directly convey impressions of outcrops to discussion partners who are unfamiliar with them. Even in a cartoon-like presentation of how structures develop, drawings are well suited. In any case, the apparatus of geological drawing must be mastered in order to draw accurately and powerfully.

In microscopy, drawings serve as a memory aid. They help record the otherwise fleeting impressions that arise during the examination of thin sections. Here, in contrast to laborious photography, drawing is an effective form of documentation. It does, however, require its own style that is different from the style of outcrop, sample, or exact thin section drawing.

Even the processing of data is facilitated by drawing. If one wants to evaluate, on the basis of one's field book, how often and where certain geological structures (e.g., foliations or shear zones) occur in a given area, even just rapidly flipping through the drawings in a field book can give a rough idea. Searching for this information in texts can be extremely time-consuming. Additionally, the close connection between drawing and text (labeling) increases the informational content. It is even possible to obtain certain information (e.g., spatial relationships of lineations to foliation planes, the orientation of joints to layering) from careful drawings that one did not pay attention to or write down in the field.

Perhaps the most outstanding feature of drawings is that they are an important tool for thinking (Larkin & Simon, 1987). Experiments provide evidence of the "analog," that is: the visual processing of information in the brain (Brooks, 1968, Shepard & Metzler, 1971—cited in Metzig & Schuster, 1993). "One technique for avoiding the rigidity of words is to think in terms of visual images and not use words at all. It is perfectly possible to think coherently in this way … The visual language of thought makes use of lines, diagrams, colors, graphs and many other devices to illustrate relationships that would be very cumbersome to describe in ordinary language" (De Bono, 1990, p. 73).

Drawing, like writing, is suitable—in many cases even more so!—for sorting and bundling thoughts. While drawing, the unimportant can be separated from the important, new ideas can form, and old ideas can be specified. It can also be resolved whether ideas are even usable and can be translated into logical and consistent models. Therefore, we should draw as often as possible while discussing, explaining, or even while thinking; what we are talking or thinking about should always be conceptualized with simple sketches or accompanied by schematic drawings. Such drawings can stretch our imagination (and that of others) and help to find the mental entry into a topic.

Finally, the circumstantial information can be more effectively saved with the help of drawings. "Pictorially presented material or visual representations [can] be stored particularly easily and permanently," and "descriptions of many creative thinkers specifically [reference] the course of pictorial ideas and the resulting creative act (like Poincaree, Kekule, Heisenberg, and others)" (Metzig & Schuster, 1993). The advantage of the schematic sketch versus an artistic, detailed, and extensive drawing has also been highlighted in investigations. "And so the development of details compared to outlined images leads not to an improvement of memory (e.g., Angin & Levie 1985). In fact, insignificant image details are forgotten shortly after being presented" (Rock *et al.*, 1972—cited in Metzer & Schuster, 1993). Although the brain is well suited for processing visual information, a photographic reproduction of individual details is not saved, "but rather, stored visual prototypes are resorted to for a mental image …" (Metzig & Schuster, 1993).

◼ Exercise 1.1:

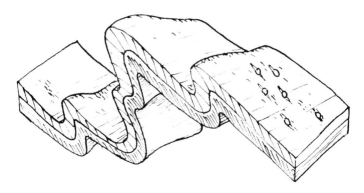

Exercise 1.1 (a) *Describe the fold structure in all of its geologically relevant details. **(b)** Give your description of the folds to a fellow student or colleague and have them draw the fold. Compare this drawing to the supplied sketch of the fold.*

Notes on Exercise 1.1:

What is geologically relevant? Certainly the monoclinic form of the fold. It refers to a movement of the "top" to the left (in the sketch). Furthermore, parts of the sketch and their geometries can be linked with geological structures and processes as follows:

- The angle of ~50-70° between the fold limbs with a "moderate" shortening,
- the "light" layering with its fan-shaped, wide-spaced foliation planes with a more quartz-rich, mica-poor composition,
- the "dark" layering with its slightly pile-shaped, narrow-spaced foliation planes with mica-rich composition,
- the foliation-free upper margin of the lower layer with a grading (mica-free versus mica-rich) and the resulting "stratigraphic up" that points, geographically, down,
- the slightly thinned fold limbs and thickened fold crests, also in the quartz-rich layer, with crystal-plastic deformation of the quartz and, consequently, with deformation temperatures of >300°C,
- the slightly curved shape of the fold axes with clearly crystal-plastic behavior of the rock layers.

10

- Additionally, the lineation on the layering (mica and feldspar- and garnet-blasts with preferred orientations) attests to the fact that (i) the foliation of the first deformation event (D1) already lies parallel to it, and, with that, the fold already represents a second fold and the sketched foliation a second foliation, and (ii) the extension of the first deformation event (D1) lies approximately perpendicular to the D2-fold axes, and that, therefore, the kinematic system during D1 and D2 was probably oriented the same way.

Probably, you will notice a difference between the present fold drawing and the fold drawing based on the description.

1.2 The Tools

When drawing in the field, the only drawing materials we need are a pen and paper. It is important that the pen leaves clean lines. The paper should be smooth (but not too smooth) and blank. Smooth paper has poor moisture retention, which gives a clean line. If the paper is too smooth, however, the pen will slip easily. Lined or graph paper interferes with drawing and scanning. Of all utensils used for artistic drawing, the ones that are most practical for geological drawing in the field are pencils, felt-tip pens, and ballpoint pens. The pencil is often recommended for drawing in the field. I, however, find it not particularly useful in most cases. If one doesn't want to keep sharpening it to avoid thick, inaccurate strokes, one must work with a hard pencil. This, however, generates pale, low-contrast drawings that require great effort by the brain to grasp and decrypt. Even when rapidly flipping through a field book, the higher-contrast drawings can be identified and compared more quickly.

A black ballpoint pen or a waterproof felt-tipped pen with a thin line offers better services compared to a pencil. Ballpoint pens also have the advantage that, depending on the pressure, lines of different weights can be generated without having to continuously sharpen the pen. The lines of waterproof felt-tipped pens can bleed on rough or wet paper. Waterproof paper is expensive or only available in field books ("rite-in-the-rain") whose layouts aren't always favorable for drawing. Damp paper, however, can be avoided with a large umbrella (the most important piece of equipment for any field geologist), for example. But if it's pouring, field work beyond just mapping is not useful, because fine structures are poorly visible in wet rocks. It is better to stay at the inn, prepare for the next field day, evaluate the drawings from the previous days, or enjoy some of the local culinary specialties. Not being able to simply erase strokes also breeds precision and thinking ahead. The pencil tempts especially beginners to draw blurredly to conceal an inaccurate observation with hatchings and the like, or to indulge their own artistic talent. Even the knowledge that "wrong" lines can be deleted leads to unconcentrated work.

Although it rarely rains during microscopy, the above applies to fast, documentary thin section sketches as well. Thin and high-contrast ballpoint or felt-tipped pen lines increase readability. One is forced to observe closely and concentrate while drawing, while abstaining from making the drawing "artistically" diffuse.

However, drawings that are meant to appear in a paper or publication and need to be precise are better executed—best with a field or thin section sketch as reference—on a new sheet of paper and in pencil first. Here, it makes sense to keep all correction possibilities open. Such drawings should be copied neatly with an ink pen that guarantees a defined and constant line weight. To get a clean and detailed drawing, it is still best to work with ink on large tracing paper. Tracing paper that is not too thin, allows for precise lines that can be corrected cleanly. Any remaining inaccuracies will disappear during digital downsizing.

Executing drawings using drawing programs is particularly popular. What remains if we disregard that fact that such programs are a wonderful toy for people who have too much time? With the software and hardware available to the average person today, large-scale and detailed drawing is not satisfactory or can only be done with great effort—also because only few people actually know how to professionally use drawing programs. Generally, the expended effort is much higher, for the first draft as well as the corrections, than with manual drawing. Drawing on the computer forces schematization more strongly and excessively strengthens the "symbolic" side of drawing (Chapter 1.3). With that, digital drawing is not useful for all the types of drawings that are discussed in this book, namely thin section, sample, and outcrop drawings, as well as geological stereograms.

However, these programs can still be used wherever schematic information is important, like in bar graphs or maps that can be kept relatively schematic, or for drawings that are to be used repeatedly as templates, for example. These programs are also well-suited for the schematic development of manually done drawings (Chapter 2.7). It is useful to scan drawings done by hand and outfit them, with the help of appropriate imaging software, with everything that can better be created schematically (labels, north arrows, scales bars, etc.). The correction possibilities of such programs should also be used. Even other functions, like contrast enhancement, line broadening or line thinning, and sharpening of lines should certainly be put to good use.

1.3   Sizes of Drawings

The size of a geological drawing, be it a quickly drawn outcrop sketch, a thin section sketch, or a drawing meticulously created for a publication, is limited both by its top and bottom borders. The simpler a drawing and the fewer details it contains, the smaller it can be. The sketch of a gneiss, which contains, aside from some foliation, a few more feldspar blasts, doesn't need to be larger than about

6×4 cm. Sketches of rocks that contain significantly more structures or composite block images and complexly structured thin section drawings require an area of at least 15×20 cm (i.e., A5)—in some cases even more.

A line cannot be arbitrarily thin and the drawing paper not arbitrarily large. Regardless of whether an ink pen, a ballpoint pen, or a pencil, 0.1 mm represents the lower limit of generatable line weights. Since the line weights in a drawing should vary to increase the readability and interpretability of a drawing (Chapters 2.4 and 3.3), the line weights are usually significantly higher than 0.1mm. However, details in a small drawing cannot be brought out by thick lines. Therefore, line weights limit the size of the drawing.

Of course, lower line weights can be achieved in print (on paper or digital). In the literature, great examples of complex stereograms can be found that have been downsized from square-meter-sized templates to areas of A3 or even A4 formats (Chapter 5). They contain lines well below the 0.1 mm line weight. But these are exceptions that are not of importance for everyday drawing in the field or at the microscope.

For thin section drawings, areas of about 15×20 cm (A5) to about 20×30 cm (A4) are useful and practical formats. The same applies to the "final draft" of sample or outcrop drawings and, in my experience, for drawings done in the field. There, the drawings often develop gradually. Progressive observations are "built on" and added to the initial drawing but the add-on direction cannot always be selected (Chapter 4). This means: The field book should be about 15×20 cm (A5) big (or slightly smaller). This allows for "standard drawings" up to A5-size and offers the possibility or extending the drawing to A4 if need be.

Larger field books are impractical, because they usually can't be stowed in jacket pockets or belt bags. I consider A6-sized field books, which only offer a double-sided drawing area of about A5, but are distributed in this size as "official" geology field books, to be too small. In combination with a thick pencil, detailed and precise drawing is no longer possible.

1.4 Geological Versus Artistic Drawing

There are similarities and differences between the two types of drawing. Neither artistic nor geological drawing is concerned with reproducing objects exactly. However, both require the ability (i) to observe, (ii) to keep what is being observed in focus while constantly switching between object and drawing, and (iii) to interpret what is being observed. Drawing has nothing to do with photographic reproduction, otherwise it wouldn't have survived the invention of photography and its development to the present state of digital processing. This, of course, does not affect that fact that every photograph is an interpretation.

Still, the differences between artistic and geological drawing are significant. Geological drawing might be easier for someone who is good at artistic drawing, although it could certainly be hampered by this ability as well. Betty Edwards (1979) explains why this is so. She elaborates on the problematic process of learning to draw and offers many useful exercises and instructions for self-study.

Up until the age of 10 to 12 years, children draw stick figures, houses made of lines, and the sun as a circle with a wreath of lines. When drawing, they use symbols for every object they are trying to represent. When they get older, are talented, or enroll in good art classes, they start to represent things more accurately. This is what is meant by "artistic drawing." Outlines of objects are no longer represented through lines or transformed into symbols; instead, a "photographic" picture is drawn that emerges through light and shadow. But most people remain at the stick figure-level for their entire life—not because they are lacking talent, but because no one ever showed them a different way of drawing.

Through language, we are trained to classify things and impose names. Similarly, we have saved a symbol for every object or class of objects. If we draw a table or a human nose, we draw the corresponding symbol. We have to classify our environment in this way, so that our thinking can work quickly and effectively. But if we want to portray things exactly, we must proceed differently. We must try to observe as accurately as possible. This is hard, requires time, and has to be learned. But there are many good books on this subject (Edwards, 1979, Sale & Betty, 2007, Jenny, 2012, et al.) and many websites with online courses.

Artistic, representational drawing works with gradations of light and shade. A constant exchange of glances between the object and the drawing is how our sketch grows. Guiding the drawing hand is as important as the concentration and rhythm that develops and contributes to silencing our logically thinking left brain hemisphere, which is always looking to impose a range of symbols on us.

Even when we draw geologically, we observe closely and try to represent an object or an image accurately on paper. But, at the same time, we must filter the geologically relevant and omit the unimportant information. This means, when we are drawing a geological structure, we are already interpreting it. We need to interpret. First, there is never enough time to represent everything in every detail. Secondly, essential elements must not drown in the ocean of trivialities. Thirdly, drawings are meant to document. This only works when schemata are used while drawing and structures across many drawings can be compared with one another. Geological drawing must therefore meet the demand of representing structures as precisely, but also as schematically, as possible, highlighting essential information, omitting the unimportant, and doing this all in the shortest time possible. This means: To make our sketch geologically understandable, we must also use symbols. We do not draw the soft, monoclinic folded rock layer the way we see it; instead, we draw the symbol of a soft, monoclinic fold in its place. The symbol must be such that it contains all the important information.

What is the important information? This is decided before we start drawing. First, we must carefully look at the geological structure and separate the geologically important from the unimportant, which means interpreting the structure. We must first understand what happened geologically, and then we draw. This, of course, means that we only recognize and draw what we can interpret. Fortunately, the situation is not quite as bad. Experience shows that one can definitely draw things without understanding them, that one can interpret while drawing, and that one instinctively retains supposedly unessential information in a drawing that later turns out to be important. Normally, however, we (first) only interpret: "Here is a foliation, and, between the planes, the folds of a former foliation can be seen," and then turn this observation into a mixture of lifelike and symbolic drawings. On the one hand, we use symbols while drawing for everything that we have geologically understood; on the other hand, we must portray some things as realistically as possible, since geological structures are often too complex and diverse for there to be a different symbol for each one. We are constantly balancing on the border between realistic and symbolic representation. This makes geological drawing quite challenging in certain respects.

1.5 Drawing With Symbols

Symbolic drawing is widespread in geology—just glance at a textbook. When we draw symbols, we represent a certain structure in a reduced schema that is immediately clear and comprehensible for everyone. This means: Symbols can only be used for well-known structures, objects, or simple processes and should only be composed of a few prominent lines. Our everyday life is interspersed with symbols. *Pictograms* are their best-known form and *droodles* perhaps the more obscure one. Even ancient pictographic scripts, like hieroglyphs, for example, relay information in the form of symbols that sometimes closely model the objects they represent and sometimes abstract significantly from them.

Strictly speaking, droodles are inverted pictograms. The search engine, Google, yields about 52,900 results (as of 08/17/2016) when the keyword "droodle" is searched. Wikipedia defines the term as, "...a picture puzzle, in which what is represented must be deciphered; the view is often an unusual or extreme perspective or an extreme outcrop." This means, droodles should represent complex things using simple shapes but should do so in a disguised, rather than in an obvious, way. They are puzzle games that feed on our addiction to and enjoyment of abstractions. A few well-known droodles are depicted in Figure 1.4 including some gems I experienced in my youth. There are usually several different solutions to each droodle. A part of their charm comes from the fact that of all the existing solutions, one can usually still come up with one's own personal favorite. However, it should not be forgotten that some people (unlike the author) simply find droodles silly.

Even writing has evolved from symbolic sketches. Especially in Egyptian hieroglyphics, the two opposed types of symbols can be found—those that stay very close to the object they are symbolizing, and those that are abstract and strongly

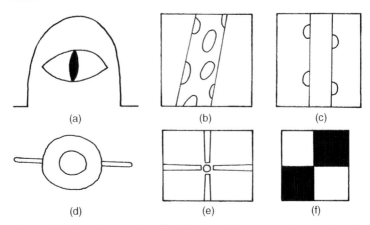

Figure 1.4 *Six well-known droodles. Together with other ones, they can be easily found in the internet. (a) Cat that peers into a mousehole; (b) giraffe standing in front of a window; (c) bear climbing a tree; (d) person with a large sun hat on a bicycle; (e) four elephants nosing at a ping-pong ball; (f) chessboard for beginners.*

deviate. Moreover, each hieroglyph also represents a letter or a number, which means that the step towards complete abstraction was already completed in this millennia old font.

Let's start strengthening our ability to understand and draw symbols. What better way to begin than with pictograms? This term originates from the Latin *pingere/pictum* (= to paint) and refers to a "single symbol or icon that conveys information through a simplified graphical representation." As symbolic terms, pictograms guide us through many situations in our everyday life—whether as go/no-go symbols in the public sphere, street signs, flight arrows, male and female stick figures, the dot in the circle that leads us to the city center, the beer mug on the hiking map that shows us the way to the nearest beer garden, the smarticon bar on our computer screen, and emoticons. These are all taken up directly by our brain, often without further thought, and processed like oral or written notifications. Pictograms are constructed with as few lines as possible, rarely contain irregularly curved lines, and are therefore well suited for computer presentations. In geology, pictograms are mostly used to portray structures and shapes, but rarely geological processes. Geological pictograms are usually not quite as simple as the pictograms in everyday life. They are only used for

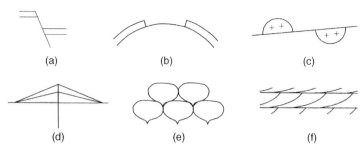

Figure 1.5 *Geological 2D pictograms ranging from kilometer to decimeter scale, reduced to the essentials. (**a**) Detachment; (**b**) metamorphic core complex; (**c**) strike-slip fault that displaces two halves of a granite pluton; (**d**) stratovolcano; (**d**) pillow lava; (**f**) cross-bedding.*

well-defined geological structures or bodies (Figure 1.5). As with the symbols in everyday life, geological symbols with no direct relation to the displayed object have established themselves. They are derived only from indirect references or have been defined on the basis of general agreement, like the symbols ⊙ and ⊗, which represent movements that are directed either towards or away from the viewer. Significant geological processes can sometimes also be represented by a series of pictograms or, in some cases, even a single pictogram (Figure 1.6).

Many structures are too complicated for pictograms. Their simplicity does not suffice to represent even just the most important components of the structure. In this case, a schematic drawing is necessary. The transition from a geological pictogram to a schematic drawing can be very smooth. These schematic representations, however, are generally structured very differently and are generated in other ways. The fact that many geological forms are scale-independent is problematic for pictograms. In contrast to the drawing of a house for example, the size of fold cannot be discerned from a drawing. Moreover, a "fold-pictogram" can only represent the type of fold but cannot differentiate between a micro fold in thin section and a kilometer large fold. Such a distinction is no problem in a photograph of the terrain, in which one can place a coin or a hammer on the rocks, or kindly ask a friendly fellow to stand against the outcrop face as a benchmark. "Fold-pictograms" and all other pictograms of geological structures must contain an additional scale to allow for discrimination based on size. This contradicts the intention of a "minimalistic" pictogram. In contrast, in geological maps, existing structures can be represented with pictograms because the scale of the structures does not matter if it's only about the fundamental representation.

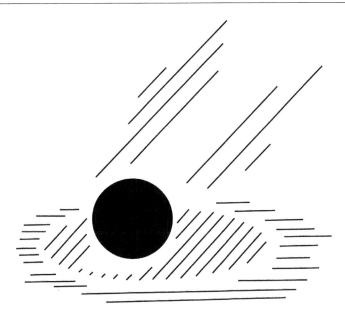

Figure 1.6 *Pictogram of a meteorite impact; redrawn symbol from a sign-board of the geopark "Nördlinger Ries" (Germany).*

Figure 1.7 *Schematic representation of rock fabrics.* **(a)** *Granitic (top) and volcanic vein (bottom) in a layered gneiss. Crosses represent feldspar cleavage planes in granite. "v" stands for volcanic rock. The homogeneous distribution of symbols represents the homogeneity of the rock fabric.* **(b)** *Orthogneiss layer in schist. The "wave" ~ symbolizes the wavy structure of the gneiss, which originates from the lenticular shapes of deformed feldspars and the distribution of the surrounding biotite flakes. The schematized fabric includes the information that the gneiss was deformed at temperatures high enough for crystal-plastic deformation of feldspar. The parallel lines represent the foliation in the schist. In addition, they contribute to the light-dark contrast that typically exists between orthogneiss and schist.* **(c)** *Bedded limestone. The cross-strokes symbolize fractures that form perpendicular to bedding during diagenesis and compaction and are rarely present in other rocks in such formation.* **(d)** *Folded limestone layer with schematized fabric adapted to the form. The cross-fractures were formed prior to folding and, consequently, remain perpendicular to the layering after folding. Watch out! This is an interpretation that needs to be supported by observation. Fractures generated during or subsequently to folding can be oriented differently.* **(e)** *Folded limestone layer with*

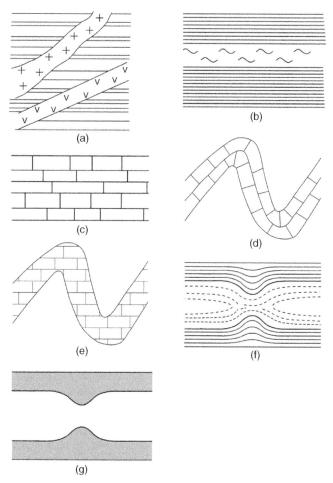

Figure 1.7 *(continued) schematized fabric not adapted to the form. Geometry of the fold and orientation of the schematized fabric communicate different messages. Their contrast may confuse viewers. (f) Boudinaged layer. Solid and broken lines represent foliation planes, which are curved as a result of boudinage. The light-dark contrast emphasizes layers of different composition. In general, the impression is generated that a strong gneiss layer is boudinaged in a mantle of weak schist. (g) The computer-generated gray-white gradation preserves the "correct" light-dark contrast; however, the foliation geometry cannot be represented.*

■ Exercise 1.2:

Draw pictograms of the following geological structures and processes: (a) graben, (b) unconformity, (c) subduction, (d) conglomerate, (e) porphyritic granite.

Notes on Exercise 1.2:

You will notice that pictograms (a) to (c) can be drawn more easily and clearly, but that pictograms (d) and (e) are less straightforward. The reason is that structures (a) to (c) have clear boundaries and are thus defined by lines, and (d) and (e) are not. The representation of a conglomerate requires a layer with boundaries as additional information, and even the pictogram of the porphyritic granite can only exist with boundaries. This is not specified for granite á priori and therefore meaningless. Furthermore, the sizes and orientations of the pebbles and feldspar crystals must be varied. Such additional information weakens the visual impact of the pictogram.

In the drawings of geological bodies, symbols, and specifically, rock symbols bridge the gap between the realistic and the schematic representation. They allow us to draw quickly and concisely under certain conditions. It is important that one does not, as is often the case with computer programs, schematically fill an area but, rather, adapts the schemas to geological formations (Figure 1.7). Geological forms are often anisotropic. This means, in different directions they show different lengths or are structured differently, for example. Symbols can also be anisotropic. This means, they too have different structure in different directions. It is bothersome when the anisotropy of the geological form differs from that of the symbol. This is mainly because the symbols are often modeled on the geological structures (Figures 1.7a-c), and deviations would contradict the "geological concept" and complicate recognition. Despite their schematization, symbols customized to the external form or the internal structure still appear natural and contain technical information (Figures 1.7d and f). Ill-adapted symbols, however, create a visual conflict and confuse the viewer (Figure 1.7e). Neutral, isotropic symbols (Figure 1.7g) are better, because they can easily be generated with a drawing program. However, geological information is lost, and they do not achieve nearly the same naturalness and inner tension as customized symbols.

Symbols are important, because they are needed for drawn rock sections regardless of their scale. Since we can never draw everything exactly as it is, we must schematize. A cleverly chosen and customized symbol allows us to bridge the gap to geological reality. Especially if we want to customize symbols, the computer won't be able to help us. Even the best drawing program is a cumbersome instrument compared to the loosely sketching hand. We will deal with choosing the appropriate symbols for different applications and how to implement them skill-fully more extensively in later chapters. Although certain rocks (or geological structures) in the geological literature have their own standard symbols, some things are left up to one's own creativity.

1.6 Realistic Drawing

The bottom line when drawing is observation and not the movement of the hand. If drawing, like any other artistic task, is to succeed, we must be in a state of concentration and suspense. Let's observe this in the next few exercises that can be found in this or similar form in many other drawing books.

■ Exercise 1.3:

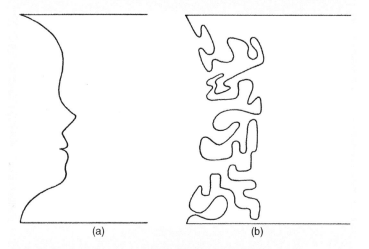

(a)   (b)

Exercise 1.3 Draw the "face" (*1.3a*) and then the abstract and more complex line (*1.3b*) mirrored between the two line ends.

Notes on Exercise 1.3:

Draw the lines slowly, at a constant speed, and in one go. In order to do that, you must continually and quickly move your eyes back and forth between the existing and the emerging line. If you do this correctly, you will find yourself in a state of concentration and suspense, which leaves no room for rambling thoughts. The better you succeed in doing this, the better your lines will be.

We draw the first given line in its mirror image—slowly, millimeter by millimeter. When drawing, the eye is constantly jumping back and forth between the original and the new sketch. We constantly monitor the progress of the line, its shape, and its distance from the mirror image. We must get into a rhythm when we observe piecewise—execute, observe—execute. When we have drawn the first, simple line in its mirror image, we move on to the second, more complex line. Repeat this exercise until the lines you draw really represent mirror images of the originals. It is particularly important that the proportions are correct!

■ Exercise 1.4:

Drawing contours.

Notes on Exercise 1.4:

In this exercise, we place a hand on the table and draw its contour, or outline, on a piece of paper using the other hand, but without looking at the sketch! We transfer what we see into the movement of the hand without using the eye to control this movement. We are not allowed to give into the temptation of looking at the sketch while we are drawing. If we repeat this exercise often enough, we get a feel for the movement of the drawing hand and for how observation is translated into movement. This exercise must be carried out very slowly. If we need 3 to 4 minutes to contour the hand, we are not drawing too slowly, but rather too quickly.

Betty Edwards (1996) gives a detailed guide to this and other drawing exercises that strengthen the ability to observe and to graphically represent what is observed (and not the interpretation of it!). Overall, such exercises work to repress symbolic drawing and improve the ability to draw realistically. Such exercises also form the basis for geological drawing and the book by Betty Edwards, or any other drawing book, can be a good introduction into the subject.

Building up concentration and suspense during geological drawing is not always easy; on the one hand, because the circumstances sometimes preclude those necessary for concentration, and on the other, because we must always also draw symbolically. The time constraints that sometimes plague us when drawing outdoors are no friend of concentration either. And when the wind tatters the field book and the cold causes the hands to cramp, it is hard to achieve and maintain an "artistic" suspense. Even the need to draw symbolically, can be a hindrance. If we always have to decide where and how to use symbols, what we want to omit and include from our observations, and in what form we should depict everything, these "rational" decisions continuously rip us out of the state of pure observation and destroy precisely the type of suspense that is important for geological drawing.

22

■ **Exercise 1.5:**

Exercise 1.5 *Upside-down-drawing: Children based on a line drawing by Heinrich Zille (1922). Draw the sketch first "right side up" and then upside down. Draw quickly and concentrated but without honing the details.*

Notes on Exercise 1.5:

If you perform this exercise as described and compare the two sketches, you should notice that the upside down drawing comes closer to the original. This is because the "right side up" drawing is recognized by our brains as people.

This causes us to move more easily toward interpretive, symbolic drawing and deviate more strongly from the template. "Upside down drawing," however, makes it easier to find a rhythm and concentration, because we are not distracted from the reality of the drawing.

■ **Exercise 1.6:**

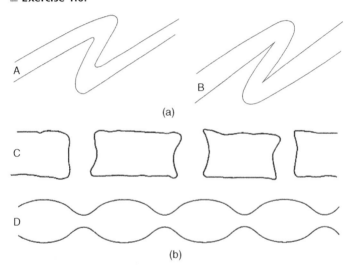

(a)

(b)

Exercise 1.6 (a) *The shape under the fold pictured in A mirrors a certain average intensity of deformation and a low strength difference between the folded layer and the surrounding rock. In comparison, the folded layer displayed under B is significantly stronger than the surrounding rock. The crests of this fold typically have a round exterior with a large radius of curvature, while the internal layer is compressed at an acute angle. When transferring natural shapes to paper, it does not usually come down to the details. The number of folds or the lengths of the limbs (although the long to short-limb ratio is usually important!), for example, must not be the exact same. Now draw, compared to (A) and (B), a folded layer with lower strength than the surrounding rock. **(b)** A boudinaged layer that is significantly harder than its surrounding is pictured under C, while the strength difference in D is less. Draw a boudinaged layer with only a slight contrast in strength to the surrounding rock.*

When we draw things that our mind has assigned symbols, we often unconsciously draw these symbols and not what we really see. It is easier to draw things that make no sense and are not composed of symbols. We draw Zille's children first "right side up" and then upside down. How are our two sketches different?

The ability to draw in a life-like manner is the basis of geological drawing. It prevents slipping into complete schematization. Nevertheless, geological drawing is influenced by schematizing—and by interpretation. What we see must be interpreted geologically. When we draw, the resemblance between the original and the drawing is less important. The determining factor is that our drawing allows for an accurate geological assertion by representing the essential elements. What is essential? Let's start with a simple exercise.

Everything that provides information about the properties of rocks and about the processes by which they were formed and changed is essential. The properties primarily include the composition of the rock and the rock fabric. Both determine not only the rock type but also influence numerous other, especially physical, properties: density, porosity, permeability, compression and shear strength, elasticity, heat capacity and heat conductivity, reactivity. In addition, a rock's history usually influences its mineralogical composition and always its fabric.

Conversely, layers in sedimentary rocks, for example, reflect different geometries and chronologies of deposition history. This is why the type of layering is an important feature of a sedimentary rock drawing. Even processes of diagenesis change the composition and fabric of a rock. Both are mainly visible in the micrometer to millimeter-scale and can be represented in thin section drawings.

Different magmatic rocks can often be differentiated from one another due to their different coarse or fine-grained groundmass. However, the graphic representation of the often diffuse magmatic structures can be very time-consuming. In some cases, however, the fabric is typical of a specific rock, like the columnar joint system of basalt, for example, and can easily be used for rock characterization.

Distinctive and clearly representable fabrics are mainly found in metamorphic rocks. Foliation planes, lineations, and folds are typical components of sample and outcrop drawings. Their chronology offers valuable information about the rock's history. Furthermore, grain fabric can be used to identify many rocks and represent them in a simple way. The microfabric, with its grain shapes and textures, offers many ways of characterizing both the rock and its history.

In the following chapters, numerous semi-schematic drawings of typical rock fabrics in the micrometer to 100-meter scale are presented and examples of how such drawings represent the history of deposition, crystallization, deformation, and metamorphism are shown.

1.7 The Fractal Geometry of Geological Fabrics

In nature, many structures are complex (fractal). They cannot, or only with great difficulty, be measured by means of Euclidean geometry but can be recorded and quantified by methods of fractal geometry (Mandelbrot, 1983). Geological structures, in particular, provide striking examples of complexity or "fractality" (Kaye, 1989), like Fe-Mn-dendrites, fracture patterns, or sutured grain boundaries, for example (Figures 1.8b-d). An essential characteristic of fractal structures is their self-similarity. Unlike mathematical fractals, that are exactly the same across an infinite number of orders of magnitude, the self-similarity of natural (including geological) fractal structures usually spans only one or two orders of magnitude (Kruhl, 2013). Moreover, this self-similarity is only of statistical nature. In addition to the fractality of structures (patterns), there is also the fractality of datasets. Grain sizes of phenocrysts in magmatic rocks or the thickness of sedimentary layers, for example, behave fractally (Figure 1.8a), meaning the size or thickness distribution follows a power law.

Both types of fractality are important for geological drawing. The statistical self-similarity of patterns means that they look almost the same in different scales and that their size cannot be deduced from their shape. This is why drawings and photographs of geological structures must always include a scale. However, this only partially applies to the schematized geological drawing of structures in specimen and outcrop scales. Since these do not create a copy of nature, but rather translate geological structures into a language of symbols, scale preservation must often be forgone in order to satisfy the symbolism. A meter thick orthogneiss in a 20 x 5 m large outcrop wall, for example, cannot be characterized through schematized feldspar lenses if these lenses are to be in the correct size ratio relative to the layer thickness (see Chapter 3.3).

If we do not copy "one-to-one," but rather, represent structures schematically and add to them, we must still observe properties that arise from the structure's fractality. If we do not, our sketches look unnatural. Above all, two properties need to be considered: (i) The self-similarity causes similar structural elements to appear in different sizes and makes the overall fabric appear "tighter" in some parts and "wider" in others. The structural elements are concentrated locally. (ii) The size distribution of structural elements according to the power law means that there are relatively few large structural parts but relatively many small ones. In a fractal size distribution of crystals in magmatic rocks, pebbles in conglomerates, or fragments in breccia, the appropriate number of small crystals, pebbles, or fragments occur and few large ones. The same goes for the distances between fractures or foliation and layer planes that are gathered closely in large numbers but are rarely further apart.

Since the breaking of rocks follows these rules, the fine structures on rock surfaces, like lineations on foliation planes, are also fractal (Figure 1.9). All of this

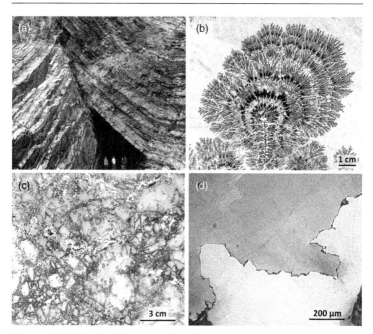

Figure 1.8 Complex (fractal) geological structures. **(a)** Devonian metapsammopelite with bedding of variable thickness; Hartland Quay (Devon, England). **(b)** Fe-Mn dendrites on a bedding plane of "Plattenkalk" (Solnhofen, Germany). The dendritic pattern is similar on different scales. **(c)** Fracture pattern on a block cut of Malm limestone; sample KR5149B; Unterwilfingen Quarry (Nördlinger Ries, Germany). The fractures are clustered, that is, form few large and many small fragments. The size distribution of these fragments follows the power law. **(d)** Photomicrograph of a sutured quartz grain boundary; sample KR4846B; syntectonically crystallized tonalite (Abbartello, Golfo de Valinco, Corsica, France). The geometrical arrangement of the few large and many small sutures follows the power law.

should be considered when drawing. Even distances between planes or linears or an even distribution of phenocrysts in rocks look unnatural compared to the distances that ensue from a fractal distribution. A sketch in which the individual structural elements are not only of the same size but spaced perfectly evenly, looks very artificial (Figure 1.10a). If the fabric is made irregular in one aspect, this only improves the overall feel slightly (Figure 1.10b). Only a variation of the fabric

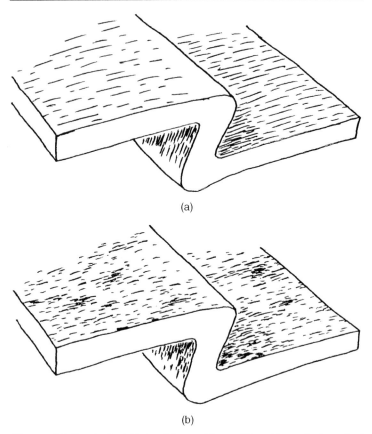

(a)

(b)

Figure 1.9 *Sketch of a folded rock layer with two different types of schematized lineations: (a) variably long and evenly distributed strokes; (b) variably long and clustered strokes. Sketch (a) looks artificial, in contrast to sketch (b).*

according to size and spacing of the structural elements along with their local concentration will yield a natural impression (Figure 1.10c).

It is not necessary, however, to elaborately construct fractal patterns when drawing. It is sufficient enough to avoid uniformity and to be mindful of including significant variations in spacing and grouping at different scales. Amazingly so, a lot can be achieved simply by "leaving gaps."

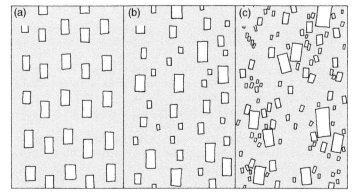

Figure 1.10 Schematic distribution (flow) pattern of feldspar phenocrysts in a porphyritic granitoid. **(a)** Crystals of equal size with equal orientation and spacing. **(b)** Crystals of different size but with equal orientation and nearly equal spacing. **(c)** Crystals of different size (few large, many small crystals) with slightly different orientations and clearly different spacing (clustering).

1.8 Basic Rules of Geological Drawing

A few basic rules for geological drawing can be derived from what has been written thus far. Although these rules will be discussed in detail in the following chapters, they should be presented in streamlined form at this point. Like all basic rules, they simply suggest the general course of direction one should follow but by no means constitute rigid commandments. One's behavior while drawing should be as variable as the types of drawings and their objectives. This is where artistic and geological drawing meet. Everyone should orient themselves according to requirements while still following their own intentions and developing their own style while sketching. Rules always include the freedom to variate or override them.

- Geological drawings should only consist of clean lines and dots. Contrasts can be reinforced and statements substantiated with different dot or line weights.
- Dots can have a geological significance but mostly serve to make surfaces different darknesses, highlight contrasts, or strengthen structures and use shadows to represent them three-dimensionally. When representing rock graininess, recrystallizations, and so on, at certain scales, dottings are especially important.

- Lines should always have a geological significance, provided they are not used to delineate artificial "blocks." Line-weight and length should be used to convey the geological message.
- Lines and dots can be used for shading as well as for geological evidence. Parallel lines cleverly distributed over curved surfaces can highlight lineations or curvatures of the surface, for example.
- Symbols may be neutral but should support, or strengthen, but never counteract, the geological information in the drawing.
- The geometry of geological structures is rarely Euclidean but usually fractal. Therefore, the naturalness of a drawing can be enhanced through grouping partial structures in different magnitudes of size and through different size distributions (according to the power law). It is usually sufficient to avoid "regularities" and leave gaps.
- A drawing should be large enough to represent all relevant structures, and details should be visible without using a magnifying glass.
- A "bad" drawing is always better than no drawing at all!

References

Anglin, G.J. and Levie, W.H. (1985). Role of visual richness in picture recognition memory. *Percept. Mot. Skills 61*, 1303–1306.

Brooks, L.R. (1968). Spatial and verbal components of the act of recall. *Canadian Journal of Psychology 22*, 349–368.

Cloos, H. (1938). Geologisch Zeichnen! *Geologische Rundschau 29*, 599–604.

De Bono, E. (1990). The use of lateral thinking, Reprint, Penguin Books, 141 pp.

Dutton, C.E. (1882). Tertiary History of the Grand Cañon District, U.S. Geological Survey, Monographs, Vol. II, Washington, 264 pp.

Edwards, B. (1979). Drawing on the right side of the brain, J.P. Tarcher Inc., 210 pp.

Einstein, A. (1955). According to Wikipedia, after eye-witness accounts cited by Franz Halla, in: Mitteilungen aus der anthroposophischen Arbeit in Deutschland, Nr. 32, 1955, S. 74–75, and by Rudolf Toepell, in: Brief an Herbert Hennig, 20.5.1955; Rudolf Steiner Archive; cited by Vögele, Der andere Rudolf Steiner, 2005, 199 f.

Hayes, B. (2008). Die Zukunft der Digitalfotografie, Spektrum der Wissenschaft 8/08, 78–83.

Heim, A. (1921). Geologie der Schweiz, Vol. II-1, Tauchnitz, Leipzig, 476 pp.

Jenny, P. (2012). Drawing Techniques (Learning to see), Princeton Architectural Press.

Kaye, B.H. (1989). A Random Walk Through Fractal Dimensions, VCH, Weinhein, 421 pp.

Kruhl, J.H. (2013). Fractal geometry techniques in the quantification of complex rock structures: a special view on scaling regimes, inhomogeneity and anisotropy. *Journal of Structural Geology 46*, 2–21.

Larkin, J. and Simon, H. (1987). Why a diagram is (sometimes) worth ten thousand words. *Cognitive Science 11*, 65–99.

Mandelbrot, B.B. (1983). The Fractal Geometry of Nature, Freeman, San Francisco.

Metzig, W. and Schuster, M. (1993). Lernen zu Lernen, 2nd edition, Springer, 257 pp.

Rock, I., Halper, F. and Clayton, T. (1972). The perception and recognition of complex figures. *Cognitive Psychology 3*, 655–673.

Sale, T. and Betti, C. (2007). Drawing: A Contemporary Approach, Wadsworth Publ./ Cengage Learning, Boston.

Shepard, R.N. and Metzler, J. (1971). Mental rotation of three-dimensional objects. *Science 171*, 701–703.

Terry, R. (1981). The art of Robert Bateman, Madison Press Books, 178 pp.

Weizäcker, C.F. von (1955). Cited after Wikipedia from eye-witness accounts of Franz Halla, in: Mitteilungen aus der anthroposophischen Arbeit in Deutschland, Nr. 32, Juni 1955, S. 74–75 and of Rudolf Toepell in a letter to Herbert Hennig, 20.5.1955; Rudolf Steiner Archiv; cited after Vögele, Der andere Rudolf Steiner, 2005, p. 199f.

Zille, H. (1922). Mein Milljöh – Neue Bilder aus dem Berlin Leben, 11th edition, Dr. Selle-Eysler A.G., Berlin, 112 pp.

Zurru, M. and Kruhl, J.H. (2000). Die Loreley. Steinalt und faltig – jung und schön! Selden & Tamm, 70 pp.

2
ROCK THIN SECTIONS

Rock thin sections provide us with a range of information that we couldn't possibly acquire with other methods requiring comparable effort. The mineral composition of a rock (and thus its petrographic classification) can only be accurately determined with the aid of thin sections. Characteristic parageneses indicate conditions of metamorphism. Above all, microfabrics supply us with information about diagenesis, deformation, and crystallization conditions. Grain boundary patterns, fracture systems, as well as the shape and distribution of crystals provide a variety of information about the conditions of formation and the rock's history. Many microfabrics that can be seen in the two-by-four centimeter world of thin rock sections influence rock structures and the behavior of rocks on larger scales.

There are various reasons why it may be worthwhile to draw microfabrics in thin sections: (i) as a form of microscopy primarily to support observations, and only secondarily for the purpose of producing a drawing; (ii) as documentation. Quick sketches made while using the microscope can later build the foundations of our records. Many things can be represented more quickly and more succinctly through sketches than in writing; (iii) as a layout drawing, with which the structure of a rock (bedding, schistosity, crystal distribution, etc.) can be visualized without addressing structure on the grain scale (intergrowth, grain boundary structures, internal fabric of crystals). Such sketches profit from the targeted omission of and focus on certain rock structures; (iv) as a schematic drawing, with which specific structures can be highlighted in a simplified form or which can be used as the basis for digitization and further processing with image analysis programs; (v) as a detail drawing, with which reaction fabrics, internal fabrics of crystals, etc. can be represented as accurately as possible.

Many, especially the older textbooks and publications, contain expressive and informative drawings of thin sections that are well worth taking a look at and reflect many of the techniques and rules in this book: Richey and Thomas (1930), Moorhouse (1959), Hatch *et al.* (1968), Mason (1978), Droop (1981), Bard (1986).

What are the advantages of a drawing over those of a photomicrograph? The simplicity of the process, for one, is a distinct advantage, especially when drawing for the purpose of fast documentation. Even with a digital camera, the light conditions must be carefully observed and painstakingly adjusted. Feldspar and quartz grain fabrics, in particular, can be rich in contrast making it difficult and time-consuming

Drawing Geological Structures, First Edition. Jörn H. Kruhl.
© 2017 John Wiley & Sons Ltd. Published 2017 by John Wiley & Sons Ltd.

to prevent the bright grains from outshining the dark ones. Furthermore, the image format options are limited and they can only be labeled and documented in hindsight. This, in turn, represents a significant advantage of the drawing, in which a detailed label is possible in a rather simple and rapid manner.

Planar structures in thin sections, such as grain boundaries, cleavage planes or surfaces of euhedral crystals, are often cut obliquely to the thin section surface and appear on photos as broad, diffuse streaks. Structures, like rutile needles found in quartz, that go deeper, are pictured out of focus. The blurring of such structures remains in the photo but can, however, be eliminated in a drawing. The ability to eliminate unimportant features and highlight important ones constitutes the decisive advantage of a drawing over a photo. Transformations of retrograde metamorphism (e.g., saussuritization in plagioclase, serpentinization in olivine and pyroxene, or pinitization of cordierite) often overprint the fabric of the rock. While corrections are hardly possible in a photo, unimportant features in a drawing can easily be eliminated and important ones highlighted. Lastly, preparation artifacts, such as trapped air bubbles or abrasive powder residues, can easily be eliminated.

The drawing does, however, pose some disadvantages to the photo. By using different interference colors or extinction orientations, certain internal structures of crystals (e.g., wavy extinction, twins, subgrain patterns, or, more generally, the different grain orientations) that were previously either not or hardly perceivable, become visible. In addition, photographing a sample is much faster than drawing it. Whenever it is primarily about fast documentation, details and precision are not important, and strong contrasts allow the structures to appear clearly in the photo, a photographic device should be used.

What information is a drawing capable of depicting? This question must be answered before we start drawing, because not everything can be drawn—firstly, because it would simply cost too much time, and secondly, because then the advantage of drawing would be lost again. This advantage lies primarily in the fact that essential information can be emphasized while unessential information can easily be omitted or pushed into the background. But what is to be documented or emphasized? When it comes to the documentation of a specific rock type, the different minerals and grain fabric must be depicted, but without the internal fabric of the crystals or minor alterations. The internal fabrics of the crystals (e.g., the subgrain fabrics of quartz, deformation twins of plagioclase, fracture patterns in garnet) are important when it comes to deformation, its strength, and its temperature dependence; the shape orientation of grains; or the relationship of individual minerals to folds and schistosity. This is how biotite flakes can, for example, be oriented in the axial plane of a fold or bent around the fold. If the conditions of metamorphism are to be highlighted, it becomes important to represent reaction fabrics in the smallest detail (e.g., finest parageneses along grain boundaries, oligoclase rims around albite, or peripheral alterations of hornblende to biotite).

When looking at all the different intentions of a drawing, it becomes clear that the thin section must largely and firstly be drawn with plane-polarized light, in which mainly grain boundaries, cleavages, chagrin, and relief are visible. Nevertheless, it may be useful to add more structures, like twins or subgrain boundaries, to the drawing (that are only visible through the use of crossed polarizers). Thus, it is possible to combine the features of two photos into one drawing.

2.1 Drawing as a Form of Microscopy

While drawing thin sections, we are primarily transferring the micro interfaces of the rock onto paper. We tend to draw more closely, leave out less, and interpret less than if were to draw rock samples or outcrops. In order to draw a detailed replication, we are forced to look at the structures in the thin section carefully and for extended periods of time ("guided vision"). We must also move our eyes quickly from the thin section image to our drawing and back. If we concentrate, this rhythmic back-and-forth brings us into a state of suspense similar to that experienced during "artistic" drawing. In this respect, sketching while using the microscope is closely related to artistic drawing.

The rapid back-and-forth from the image in the microscope to the drawing sheet may cause problems, because the eyes constantly have to refocus. For microscopy, they are set to infinity (when the microscope is set correctly), but while drawing, the focus is approximately 30 cm from the eyes. Binoculars force the head to move back and forth, which could cause a collision between nose and eyepiece. Therefore, it is advantageous to work with a monocular microscope. With one eye looking through the eyepiece and the other onto the paper, the head need not be constantly repositioned and sketching can proceed in a swift and relaxed manner. This can of course also be done with a binocular microscope, where only one "half" is used. However, in one-eyed microscopy, not every eyepiece needs to be separately focused on the eye's visual acuity and the eyes can be switched when signs of fatigue appear.

Drawing while using a microscope has the advantage that it forces us to take long and exact looks at parts of a thin section. Before we can draw what we see, we must be aware of the shapes and structures. This is how drawing allows us to get a better insight into the details of the rock's fabric.

2.2 Drawing with Various Tools

In times of digital photography and diverse printing options, thin sections can quickly be made into *paper photos*, from which the thin section image can then be transferred to tracing paper. If high precision is not required, one can easily obtain a rough sketch with the same proportions as the original. This method is fast and (usually) results in sufficiently precise drawings. Details of the grain fabric and internal structures of the crystals, however, are sometimes not depicted in enough detail in the photomicrograph. So, if precision and details are required, the sketch must be elaborately corrected and details added with help of the microscope image. Being confined to the frame specified by the photo and, most importantly, that the photo can only be taken with the polarizers in a specific position (unless one wishes to take multiple photos with different polarizer positions as is done, for example, in the automatic detection of phase boundaries—Fueten & Mason 2007) are further disadvantages of this method. Thus, one must weigh the advantages and disadvantages of parallel versus crossed polarizers, for example.

Of course one can also remain completely in the digital domain. Digital drawing versus the digital image has the advantage that one can draw in high resolution while preserving high image quality. This guarantees precision, but requires a lot of time, and, when drawing on a larger scale, a steady hand is needed and some details inevitably remain angular. At best, they look unnatural; at worst, they may even be falsified. Advantageous, however, is the fact that grain and phase boundaries, for example, can be created in different planes. This separation allows for separate processing. This is particularly useful if the fabric is to be quantified by means of image analysis programs.

In the age of digital image capture and image processing, the *drawing tube* has gone out of fashion. It is a shame as its advantages are obvious. As with a free sketch, the frame and magnification are generally freely selectable; unimportant details may be omitted and important ones emphasized. The alternating use of parallel and crossed polarizers allow for the microstructures in the thin section to be made more visible and, thus, to be sketched more easily. Angles between the polarizers, which deviate from 90°, have the same effect. The proportions in such "one-to-one" drawings are also preserved.

The main advantage of the drawing tube is that, with its help, drawings can be made not only in high resolution but also in (almost) any size. This is achieved through controlled movement of the thin section on the microscope stage in two directions perpendicular to each other, along with the use of custom-sized drawing paper. Even large sheets of paper can be pushed back and forth under the drawing tube by rolling up the sides. Or, smaller paper can be used and later glued together. Minor technical difficulties can be overcome easily. Especially the luminous intensity of the microscope and the illumination of the paper must be precisely matched. Aside from the fact that a drawing tube is no longer present at every institution, the time-consuming nature of this device alone poses a disadvantage. Its use is therefore only useful if very accurate and/or large-format sketches are necessary.

2.3 Foundations of Thin-Section Drawing

For the creation of "contradiction-free" and "technically-clean" drawings, it is important to keep in mind the way in which rock and mineral fabrics appear in thin sections. (i) Thin sections virtually represent two-dimensional cross sections of three-dimensional fabrics. This means that 3D crystal shapes in thin section often appear in characteristic 2D sections. (ii) Certain minerals possess special properties, such as refraction (that are expressed in relief and chagrin) or internal fabric (cleavages, subgrains, twins, etc.), which can be used for their characterization. (iii) Which mineral or crystalline properties should logically be depicted, is determined by the scale of the drawing. (iv) Because our brain is able to supplement gaps and piece together complete images from incomplete ones, thin-section drawings need (should) not be "complete," but may (should) contain gaps—to periodically relieve the observer and to save time.

2.3.1   Thin sections as 2D sections through a 3D fabric

Every cross section through a sphere results in a circle, whereas cross sections through a cube can differ significantly from one another in terms of their outline and represent triangles, trapezoids, rectangles, or squares. Assuming that minerals develop eu- or at least subhedrally, they form certain sections in thin section that can be used to characterize them (Figure 2.1). Even when a mineral occurs anhedrally, its cross section can still be characteristic. Columnar minerals (sillimanite, apatite, rutile) remain stretched in sections or occur with characteristic rounded, usually even euhedral cross sections. Even flaky or tabular minerals (e.g., mica) generally have stretched sections with straight, longitudinal edges. Garnet cross sections are predominantly rounded. Even for schematic sketches, the 3D-2D relationship must be considered for the "correct" cross section can convey, even without laborious labeling, important information about crystal orientation and mineral species. In labeled sketches, it can be very irritating when minerals possess the "wrong" form. Several textbooks are available which show photomicrographs with euhedral crystals of the most common rock-forming minerals and provide a good overview on the variation of crystal shapes (MacKenzie & Adams, 1994, Perkins & Henke, 2003, Raith *et al.*, 2012).

2.3.2   Specific mineral properties

The refraction of light is an essential characteristic through which a grain's *relief* and *chagrin* are determined in thin section. The term *chagrin* (French), a certain type of leather from horse or donkey hides with a rough surface, refers to the internal texture of the grain. In preparing the thin section, the grinding process does not smooth the grain surface entirely. The uneven surface is filled with the adhesive used to secure the cover slip (usually a resin with a refractive index of $n \sim 1.56$). Because the refractive index of the adhesive and the grain rarely coincide exactly, "Becke lines" form at the boundaries of those with adhesive-filled indentations and create a network of fine light lines and light spots (so-called "chagrin"). The larger the difference in refraction between adhesive and mineral (especially large in minerals with high refraction), the higher the chagrin.

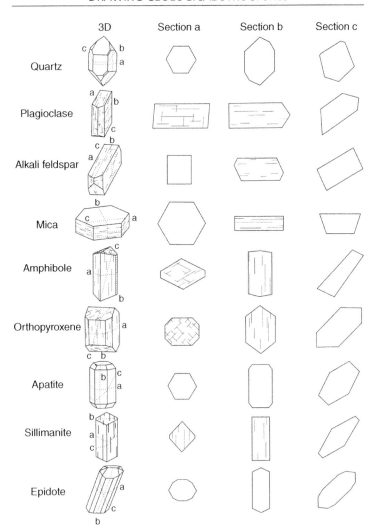

Figure 2.1 *Crystal shapes and characteristic sections of some rock-forming minerals. In addition to the 3D form, transverse and longitudinal sections, as well as an arbitrary section, are shown. 3D forms modified after Tröger* et al. *(1979); see text for details.*

Relief occurs when grains of different refractions lie adjacent to one another. The surfaces of grains with high refraction seem to be higher than the surfaces of grains with lower refractions—they have a "high" or "low" relief. This effect is caused both by a difference in chagrin and by the fact that the Becke line in minerals with high refraction is more strongly developed than in minerals with low refraction. In a drawing, the optical impression of a high relief is achieved through relatively wide grain boundary lines, occasionally with added dotted lines around the edges. High chagrin is modeled through thick dotting (or rather: fine rings) in the grain.

The internal fabric refers to cleavages, subgrain structure, twins, fractures, inclusions, and so on. These may be characteristic of certain minerals. For example, *cleavage planes* in pyroxene and amphibole are particularly well developed in two specific orientations. In pyroxene, they form an angle of approximately 87° and, in amphiboles, one of approximately 56°. However, the angle reaches these values only in sections perpendicular to both cleavage planes, that is, perpendicular to the long axis of pyroxene and amphibole; in all other sections the angle is lower. Nevertheless, the difference between the angles is so large, that it is recognizable even in deviating sections and thus can be used as a characteristic feature in a drawing. In certain minerals (e.g., garnet), "brittle" behavior develops into characteristic systems of densely accumulated and often slightly curved fractures along which the grain undergoes typical alterations. Such alterations are characteristic of cordierite (pinitization) and olivine (serpentinization, uralitization).

Twins occur in many minerals but are especially common in plagioclase, K-feldspar, and cordierite. On the one hand, this allows these minerals to be drawn in a characteristic manner, and, on the other hand, the shape of the twins reveals something about their formation conditions (deformation versus growth twins), which is why a correct drawing is important. The fact that twins are only visible through different extinctions with crossed polarizers can, however, present problems while drawing. For example, since the refraction of light of individual twins in feldspar is only slightly different, twin boundaries, with the exception of calcite, are hardly visible in plane-polarized light.

Even *subgrains* may occur in a characteristic form in certain minerals. Relatively high temperatures are indicated by a chessboard pattern in quartz (two systems of orthogonal subgrain boundaries), unlike parallel subgrain boundaries, which occur in the middle and lower temperature ranges of metamorphism. Subgrain boundaries are only visible through the different extinctions of adjacent crystal regions with crossed polarizers and not at all visible in plane-polarized light. It is because of this that their representation is problematic. Usually, as a help to oneself, one outlines the boundaries between regions of different extinction as dotted, and therefore, "finer," lines.

Even *reaction fabrics are* an important source of information and are generally included precisely in most drawings. Particularly in this aspect, a drawing can reveal much more than a photo. Especially structures that can provide information about the relative growth age of different phases, about mineral transformations, or about the relationship between growth/reaction and deformation, are important for the analysis of rock or material history and must be drawn both correctly and in detail. Here exists a (only apparently paradoxical) interaction between drawing and understanding: To be able to correctly draw microfabric, one must have understood it, and to understand it, one must draw it accurately.

2.3.3 Scale-dependent illustrations

Which mineral properties are depicted in a thin section drawing also depends on the scale. Large scale drawings allow fine details like the rotation of the twin boundary, which occurs when an albite twin is permeated by a pericline one; details of symplectitic intergrowth; or the fine sutures of quartz-quartz-grain boundaries to be represented. In small-scale drawings, it is advisable to schematize the internal fabric of the minerals or the fine-grained rock matrix. It always depends on being able to use the schematization to emphasize characteristic features of the fabric. Despite schematization, the basic characteristics of the fabric must be maintained and not misrepresented. Of course, even in a schematic representation, the twin lamellae of plagioclase must lie parallel (albite-twinning) and (nearly) traverse (pericline-twinning) to the flat face of euhedral crystals, and the amphibole cleavage planes should enclose the correct 56° angle. If porphyritic rock is foliated, the foliation of the matrix should be represented by fine parallel lines in the schematic drawing. Phenocrysts should be arranged with their flat faces in the foliation. The amount of deformation can also be depicted qualitatively in this kind of schematic representation. Further possibilities of schematization are presented and discussed in Section 2.4.

2.3.4 Leave gaps!

Have the courage to leave gaps! One can safely rely on the brain's ability to fill in gaps and to "see" things that are not necessarily present. Drawings are often packed too full, which can cause essential details, as well as a drawing's ease, to be lost in the throng of trivialities. Even when detailed thin-section drawings are "complete" with respect to the grain boundary pattern, reasonable editing of the internal fabric can achieve a more "natural" look and also saves a lot of time. Even if the fabric is "complete," our brains expect and actually "see" gaps. Gaps in the images that surround us every day are commonplace; the brain is accustomed to composing complete images from such incomplete patterns. In his book, "Visual Intelligence: How We Create What We See" (1998), Donald D. Hoffman describes this particular ability of the brain clearly. Not least, the Japanese art of ink painting offers impressive examples of "incomplete" images (Okamoto, 1996).

2.4   Minerals and Their Characteristic Fabrics Under the Microscope

2.4.1   Characteristic 2D shapes of rock-forming minerals

Especially in schematic sketches, but also in detailed drawings, it is an important help for the viewer if individual grains in cross sections characteristic of a certain mineral are displayed correctly. The cross sections of euhedral crystals of rock-forming minerals shown in Figure 2.1 represent only a small sample of the spectrum of possible outlines, yet they illustrate that a specific span of possible geometrical outlines exists for specific minerals.

Euhedral *quartz* usually occurs only in volcanic, and rarely in plutonic, rocks. In that case, hexagonal sections and compact variations are typical. In metamorphic rocks, quartz cannot prevail in its euhedral shape. *Plagioclase* occurs only in euhedral, mostly tabular, form in igneous rocks (Vernon, 1986) with correspondingly elongate, rectangular sections, and sometimes with trapezoidal parts. Similarly to plagioclase, alkali feldspars also occur in euhedral form in igneous rocks (Vernon, 1986), often comparatively more elongate and with corresponding rectangular or almost square sections. In white mica, biotite, and chlorite, the majority of all sections are elongate due to the pseudo-hexagonal shape. This effect is reinforced also by the fact that thin sections in metamorphic rocks are almost exclusively oriented perpendicular, and only in very rare cases parallel, to the schistosity. Mica almost always occurs in hexagonal and round sections in those sections oriented in parallel.

Amphiboles also prevail in their euhedral form in metamorphic rocks with respect to their surroundings. Sections perpendicular to the long crystal axis provide a distinctive rhombus shape with outer faces and cleavage planes encompassing angles of 56° or 124°. The longitudinal sections are generally elongate and approximately rectangular with mostly clearly visible cleavage. The euhedral forms of pyroxenes deviate in detail from one another depending on the type of pyroxene. Common to all is their short columnar form (considerably more stocky than that of amphiboles) with pronounced lateral faces of which four stand nearly perpendicular to one another. The faces at the column ends may or may not be developed. Sections across the column range from being eight-sided to being approximately square. Sections parallel to the column form compact rectangles and may have beveled edges.

Apatite is found in most igneous and metamorphic rocks, almost always in euhedral, hexagonal stalks that vary from short to long. The cross sections usually yield both equilateral and elongate hexagons. *Sillimanite* often develops its euhedral shape, slim to stocky columns, with an unmistakable, almost quadratic cross section and characteristic cleavage planes in the smaller angle. *Epidote* can take on a euhedral to subhedral form in igneous, as well as metamorphic, rocks. Its long-to short-stemmed shape leads to rhombic or oblong cross sections.

2.4.2 Fabric-based mineral identification

Since many minerals occur anhedrally or, at most, subhedrally in igneous rocks, further features, such as chagrin, relief, and the internal fabric, play a larger role in the graphic delineation of various minerals. This means: Regardless of whether or not one creates a schematic sketch or a detailed drawing, different minerals are chiefly represented through different line weights, dotting, and internal structure.

Line weight: For mineral grains of high refraction, the boundaries in thin section are relatively thick compared to mineral grains of lower refraction. Therefore, the relative line weight of grain boundaries in thin-section drawings should match the relative refraction of the corresponding mineral. For example, garnet or titanite should be assigned the heaviest line weight, amphiboles or mica one in the middle, and quartz and feldspar the lighter line weights. Strictly speaking, however, one cannot assign one mineral a specific line weight, because the weight of a grain or phase boundary (i.e., the intensity of the Becke line) in a thin section is dependent on the refraction of the two adjacent mineral grains and can further vary due to the anisotropy of refraction within a single mineral grain. For practical reasons, however, these subtleties of variations in light refraction are ignored, and every mineral is assigned (based on a drawing) a specific line weight for the boundaries. This does not include the boundaries between two grains of the same mineral; these are always depicted as relatively thin, because there is little to no Becke line.

What is a "heavy" or a "light" line weight? That depends on the size of the drawing, a possible subsequent reduction in size, and the difference between the highest and lowest refraction in thin section. After a reduction in the size of the drawing, the lightest line weight should still be clearly visible. Lines that are (too) thick, on the other hand, make the drawing look bulky and make it difficult to represent grain perimeter details. In a drawing of A4 size, it has proven to be practical to vary the line weight between 0.1 mm and 0.8 mm. If significant reductions in size are intended, these values must be increased. Although each mineral can be assigned a line weight in accordance with its (average) refraction (Figure 2.2), deviations from these values may sometimes be necessary. One should, in order to make a sketch high-contrast and, thus, clearer and easier to read, exploit, or at least extend, the range of possible line weights beyond the usual schema. This means, for a thin section that only contains quartz, plagioclase, and white mica, the line weight of white mica should be increased without, of course, clearly violating the usual gradation.

Dotting is, next to leaving the page blank, the only neutral signature in a thin-section drawing. One can use it to (i) darken surfaces, for example, to represent the inherent color of biotite or amphibole, and (ii) to enhance the relief and chagrin. To illustrate the inherent color, surfaces should be finely and densely

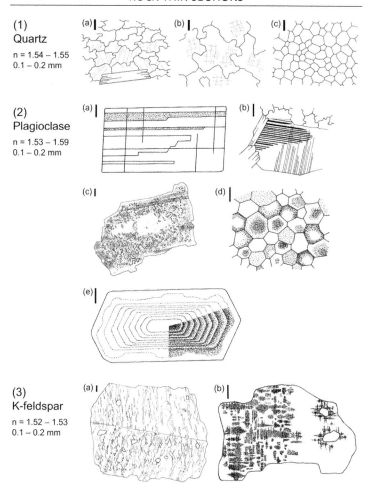

(1)
Quartz
n = 1.54 – 1.55
0.1 – 0.2 mm

(2)
Plagioclase
n = 1.53 – 1.59
0.1 – 0.2 mm

(3)
K-feldspar
n = 1.52 – 1.53
0.1 – 0.2 mm

Figure 2.2 Schematic illustrations of important rock-forming minerals with characteristic shapes and internal fabrics as observed in thin section. For each mineral, the range of refractive indices, n (after Tröger et al., 1979), and the resulting approximate range of line weights are specified. The line weights relate to sketch sizes of ca. A5 to A4. Black bars represent 1 cm in the original sketch. Characteristic internal fabrics should be used specifically in schematic sketches. The illustrated ones represent selected examples and by far do not cover the entire range of potential fabrics.

43

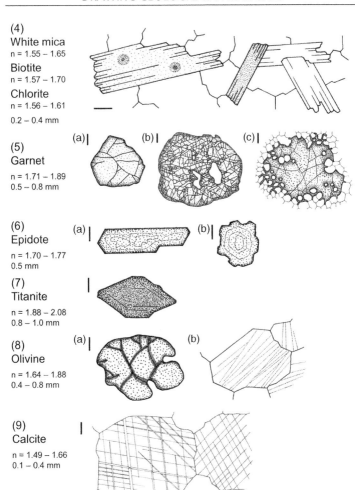

(4)
White mica
n = 1.55 – 1.65
Biotite
n = 1.57 – 1.70
Chlorite
n = 1.56 – 1.61
0.2 – 0.4 mm

(5)
Garnet
n = 1.71 – 1.89
0.5 – 0.8 mm

(a) (b) (c)

(6)
Epidote
n = 1.70 – 1.77
0.5 mm

(a) (b)

(7)
Titanite
n = 1.88 – 2.08
0.8 – 1.0 mm

(8)
Olivine
n = 1.64 – 1.88
0.4 – 0.8 mm

(a) (b)

(9)
Calcite
n = 1.49 – 1.66
0.1 – 0.4 mm

Figure 2.2 (continued)

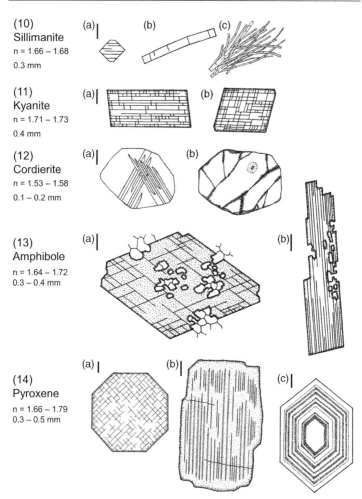

(10)
Sillimanite
n = 1.66 – 1.68
0.3 mm

(11)
Kyanite
n = 1.71 – 1.73
0.4 mm

(12)
Cordierite
n = 1.53 – 1.58
0.1 – 0.2 mm

(13)
Amphibole
n = 1.64 – 1.72
0.3 – 0.4 mm

(14)
Pyroxene
n = 1.66 – 1.79
0.3 – 0.5 mm

Figure 2.2 (continued)

dotted. Thick and widely spaced dotting in the grain interior, however, amplifies the chagrin. Grains, especially those of highly refractive minerals, are dotted in this manner. Fine circles, however, are preferable to thick dotting, but generating them is also associated with a bit more effort. The dotting inside and close to the edge of a grain reinforces, together with a thick line, the contrast between adjacent crystals, visually distinguishing the higher refractive grain from its neighbors. Through this, relief is modeled in thin section. For this reason, this dotting is not used at boundaries between grains of the same mineral or for fractures and cleavage planes.

Internal fabric: Those with experience in microscopy don't identify minerals based on their optical properties (optical character, refraction, maximum birefringence) but rather by their internal fabric or, more importantly, optical characteristics in combination with specific internal structures: lamellar twins and saussuritization in plagioclase of higher calcium content; exsolution and the characteristic microcline twinning in ordered alkali feldspar; cleavage in pyroxene and amphibole in combination with higher refraction and characteristic natural color; subgrains and (often times) sutured grain boundaries in quartz. Correspondingly important is the representation of the internal structures in the thin-section drawing. Furthermore, it is important that the internal structures be represented in correct crystallographic orientation: (i) the amphibole cleavage planes in a 56° angle to one another and the pyroxene cleavage planes in an 87° angle (except when cut effects are made where the crystal axis is significantly misaligned from the cut), (ii) the lamellar deformation twins in plagioclase parallel to the flat side of tabular grains (albite twins) or approximately perpendicular (pericline twins), (iii) subgrain boundaries in greenschist and amphibolite facies quartz as a set of parallel boundaries. Subgrain boundaries (and possibly twin boundaries) are the only structures that are shown in the drawing as finely dotted lines. This highlights that in plane-polarized light they, as opposed to grain/phase boundaries or fractures, are not or hardly recognizable. Together with the line weight, the characteristic internal fabric of different minerals (Figure 2.2) works to make each mineral in the thin section drawing easily recognizable and stand out against other minerals.

Wavy extinction, alterations, interference colors, and cut effects belong to the structures or optical characteristics that usually are not (cannot be) depicted. Wavy extinction is relatively uncharacteristic and can usually be omitted; if included, fine-scale dotting is the only way of adequately representing the diffuse light-dark shift. Even alterations (e.g., uralitization in olivine) are usually not shown, because they are not necessary for mineral characterization. The ability to elucidate important fabrics by leaving out alteration structures is a big advantage of the thin section drawing over the photo. In plagioclase, however, it can be useful to sketch saussuritization to distinguish it from quartz and alkali feldspar, and likewise as an indication of a general rock alteration. An exception is the weak (not the complete!) pinitization in cordierite. If this transformation to white mica appears only closely along fractures and grain perimeters, it constitutes an important diagnostic feature of cordierite (Figure 2.2.12b). Interference colors are of little diagnostic value in photos and cannot be represented in black and white drawings. Cut effects, especially obliquely cut grain and phase boundaries, can interfere significantly in photos. In thin section drawings, they are replaced by thin lines usually in the middle of the diffusely confined strip. The resulting small inaccuracies in the border demarcation are mostly irrelevant.

The typical sketches of rock-forming minerals depicted in Figure 2.2 illustrate how well it is possible to graphically characterize various minerals.

2.4.2.1 Quartz Due to its comparatively low refraction, quartz, along with feldspar and cordierite, is assigned the lowest relative line weight. Since quartz behaves plastically in temperatures above $\sim 300°C$, therefore under almost all conditions of metamorphism (Voll, 1976a, Vernon, 2004, Passchier & Trouw, 2005), certain deformation structures are typical of quartz in metamorphic rocks: recrystallized grains, subgrains, and grain boundary sutures. With its often well-defined subgrain boundaries and sutured grain boundaries with short, ever straight segments (Kruhl & Peternell, 2002), quartz can almost always be differentiated from other minerals (Figure 2.2.1a,b), especially from plagioclase and K-feldspar, with significant certainty. Even polygonal cellular fabric ("foam texture") with, statistically speaking, 120° angles at triple junctions, represents a characteristic quartz fabric (Figure 2.2.1c). These equilibrium angles often come about when the boundaries bend shortly before the triple junction. Even when they are shown schematically, these fabrics contain information about the temperature conditions as well as the deformation and temperature evolution of the rock. Accordingly, one should proceed carefully when drawing so as not to thoughtlessly convey false information. If quartz grain boundaries encounter basal planes of mica, they will always form a 90° angle (Figure 2.2.1a; Voll, 1960).

Although subgrain boundaries are not visible in plane-polarized light, they should be shown (dotted lines), for they help differentiate quartz from the feldspars. In addition, prism parallel and "checkered" subgrain boundaries allow quartz deformed under greenschist and amphibolite facies conditions to be distinguished from quartz deformed under granulite facies conditions (Kruhl, 1996) (Figure 2.2.1a,b).

2.4.2.2 Plagioclase Plagioclase is represented with the same line weight as quartz. Its internal fabric, however, is markedly different. Relatively wide and often tapered growth twins (Figure 2.2.2a) can easily be differentiated from the generally thinner and more pointed deformation twins (Figure 2.2.2b), which moreover appear in two sets almost perpendicular to one another. Above all, the slim deformation twins are a unique feature of plagioclase and should be used for schematic characterization of the mineral. Therefore, it is useful to sketch twins even though they are (almost) invisible in plane-polarized light. Wider twins can be dotted or not filled in (Figure 2.2.2a); with slimmer ones, a thicker pen can be used to provide a black fill (Figure 2.2.2b). The saussuritization (Figure 2.2.2c) is characteristic of plagioclase with higher calcium content. It can describe a zonal Ca-distribution in the crystal or not. Typical, in any case, is a parallel arrangement of fine mica flakes or epidote-/clinozoisite columns in which the dominant lattice planes of the crystal mirror themselves. Isometric, recrystallized grains ("foam texture") are also typical of plagioclase (Figure 2.2.2d). Its sometimes diffuse metamorphic zoning is best represented by dotting of various degrees. It differs markedly from the (often oscillating) zoning of magmatic crystals (Figure 2.2.2e), which is only one-quarter drawn in the diagram to save time. Dotted boundaries of the zones are generally sufficient to display the zoning. The oscillating zoning constitutes an important indicator of the magmatic origin of plagioclase and is therefore worth being drawn. The saussuritization typical of plagioclase should, however, be drawn only in justified cases. It is not only difficult and time-consuming to represent, it can contribute to the concealment of other features.

2.4.2.3 Alkali Feldspar

2.4.2.3 Alkali Feldspar   As with quartz and plagioclase, the use of the same line weight is appropriate for alkali feldspar (K-feldspar). The K-feldspar also has a typical internal fabric, especially exsolution and a combination of minute albite and pericline twinning (gridiron twinning). The exsolution forms regions of different geometries, often elongate lamellae (Figure 2.2.3a), but, due to their irregular boundaries, these can be readily distinguished from twin lamellae. Growth twins, according to the Carlsbad law, are common but do not constitute a diagnostic feature. Especially the lamellar regions of exsolution are crystallographically weakly oriented, that is, assume slightly different positions in the two halves of a growth twin (Figure 2.2.3a). Unlike in plagioclase, lamellar twins appear only with a diffuse boundary, "pinch and swell," and in two mutually perpendicular sets (Figure 2.2.3b) and are therefore easily distinguishable from the plagioclase twin lamellae. They form preferably in areas of increased stress, like inclusions, for example. If drawing with an ink or ballpoint pen, the typically diffuse lamellar boundaries can only be created through (time-consuming) dotting.

2.4.2.4 White Mica, Biotite, Chlorite

2.4.2.4 White Mica, Biotite, Chlorite   The crystals of these three sheet silicates are similarly shaped: thin tabular with straight (igneous or high-grade metamorphic) or ragged (low- to medium-grade metamorphic) end faces (Figure 2.2.4). Compared to quartz and feldspar, the somewhat higher refraction requires slightly wider strokes when drawing. These sheet silicates distinguish themselves from other minerals not only through their form, but also their cleavage. Especially at the ends of the plates, the cleavage planes are pronounced. The mostly green and brown inherent colors of chlorite and biotite can be represented well through a varying density of dots. Dark, pleochroic halos around small inclusions (often zircon), where nuclear radiation has partially or completely destroyed the mica lattice, are also diagnostically relevant.

2.4.2.5   Garnet   The high refraction requires bold strokes. High relief and chagrin are represented by internal dotting and dotting around the edge (Figure 2.2.5). Open "rings" (Figure 2.2.5a,c) mimic the real nature of the chagrin more closely than dots. Rounded shapes and a network of slightly curved fractures, which may be weak or intensely formed, are typical of garnet (Figures 2.2.5a,b). As the often open fractures appear dark in thin section, they are drawn with the same or only slightly lighter line weight as the outlines. Inclusions (mostly quartz) can also be a valid characteristic of garnet (Figure 2.2.5c).

2.4.2.6   Epidote   Depending on the cut, rounded or elongate forms with partially developed crystal faces are typical of the mineral (Figures 2.2.6a,b). The slight yellow-green inherent color that appears at higher levels of Fe-content can be recreated through a light dotting. The zoning is only visible with crossed polarizers, but can be sketched with a dotted line and still be used as a diagnostic feature.

2.4.2.7   Titanite   Titanite has the highest refractive index of the minerals mentioned here. This requires a thick line in the corresponding drawing. In thin section, the grains usually appear dark-brownish in color. In the drawing, this, along with the high refraction, is represented by a thick dotting (Figure 2.2.7). Even the occasionally occurring euhedral, rhomboidal diamonds are characteristic.

2.4.2.8   Olivine   Depending on the chemical composition (Fe versus Mg) and origin (magmatic versus metamorphic) the crystals may be formed significantly differently. Fe-rich olivine is prone to strong alteration (iddingsitization or uralitization), has a high refractive index, and requires a correspondingly heavy line weight in sketches. Resorption embayments (Figure 2.2.8a) are also quite typical of magmatic olivine. Fibrous serpentine often forms at the grain boundary or along fractures (which are generally curved). Similarly to quartz, magmatic olivine, like that in ultrabasic bodies in the lower continental crust usually with a lower refractive index, for example, develops subgrain patterns (Figure 2.2.8b) and sutured boundaries to neighboring grains.

2.4.2.9   Calcite   The extreme differences in refraction of calcite require, strictly speaking, a large range of line weights. But that would lead to irritating differences in calcite's graphic representation. Therefore, calcite is usually drawn with a thin line (Figure 2.2.9). Calcite's internal fabric is unmistakable in thin section: cleavage planes at an angle of 75°, deformation twin lamellae parallel to these cleavage planes, and a set of lamellae in the sharp angle between the cleavage planes. Since calcite easily reacts to stress, this fabric develops almost always, with the exception of extremely fine-grained material.

2.4.2.10   Sillimanite   The form of the mineral is characteristic and can easily be sketched. In cross section, the sillimanite columns form almost square faces with angles of 88° and 92°. The generally significant cleavage is always at the smaller angle (Figure 2.2.10a). Bunches of fibers often develop from thick columns (Figure 2.2.10c). In longitudinal section, the columns are often bent at cross fractures (Figure 2.2.10b).

2.4.2.11 Kyanite This mineral also has characteristic cross sections and internal fabric: angles of 78° and 87° between cleavage planes and between crystal faces in longitudinal sections (Figure 2.2.11a), and 74° in cross section (Figure 2.2.11b). The cleavage planes frequently do not intersect but "block" each other in their development, as is the case with other minerals that have perpendicularly stacked cleavage planes.

2.4.2.12 Cordierite Despite a light refraction (and, thus, a line weight) similarly low like that in quartz and feldspar, the typical differences between cordierite and these minerals are easy to graphically delineate. Especially the cyclical twinning, which appears as angled sets of parallel twin lamellae in thin section, are not found in any other mineral (Figure 2.2.12a). Equally typical is the alteration into tiny white mica flakes along grain boundaries and (usually curved) fractures (pinitization) (Figure 2.2.12b). As with biotite and chlorite, radioactive inclusions can cause pleochroic halos that appear yellow to light brown and are presented as finely dotted circular areas in the cordierite.

2.4.2.13 Amphibole Despite highly variable chemistry, the form and internal fabric of various amphiboles is relatively uniform: a 56° angle between the outer and cleavage planes in compact cross section, and a significantly elongate longitudinal section in which the cleavage planes are often closely spaced (Figure 2.2.13). Since amphibole usually asserts and retains its euhedral shape against its environment, euhedral sectional figures are characteristic of the mineral. The often-strong intrinsic color can be represented by dense internal dotting (Figure 2.2.13a). The narrow spacing of cleavage planes appears sufficiently dark even without dotting (Figure 2.2.13b). The common mineral inclusions (mostly quartz or plagioclase) also serve as features in amphibole, which otherwise occur in this form only in garnet.

Figure 2.3 Greenschist-facies orthogneiss from the eastern Tauern Window (Eastern Alps, Austria); sample KR2916A; mineralogical composition: quartz (Qtz), plagioclase (Pl), biotite (Bt), chlorite (Chl) and epidote (Ep); pencil drawing based on a microscope image in plane-polarized light; size of original drawing ca. A4; line weight 0.5 mm and locally lighter, due to obliquely set lead. The drawing develops successively. (a) Contouring of the larger mineral grains and the epidote layer with the same light line weight for easier correction. The end faces of the biotite platelets are sketched straight, indicating the magmatic origin of the

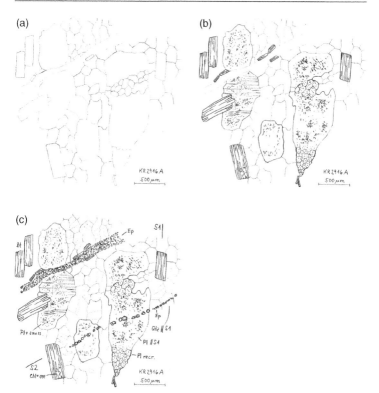

Figure 2.3 *(continued) crystals.* **(b)** *Intensification of contours and internal fabrics in plagioclase and biotite. Saussuritization (sauss) in plagioclase is schematized for the purpose of saving of time. The grain margin, poor in saussurite, is highlighted, because it reflects a magmatic zoning and, consequently, the magmatic origin of the crystals.* **(c)** *Finalization of the drawing by labeling, filling the epidote layer, drawing of the quartz subgrain boundaries, and adding the second (incomplete) epidote layer that was missing before. This relatively cursory sketch serves to illustrate the characteristics of rock deformation: (i) an early magmatic to sub-magmatic deformation, which led to foliation (S1: shape orientation of biotite, plagioclase and coarsely sutured quartz) and local recrystallization of plagioclase, and (ii) a subsequent brittle deformation that generated epidote-filled fractures (S2) oblique to S1.*

51

2.4.2.14   Pyroxene   As with amphibole, the two sets of cleavage planes are also a characteristic feature, however, in pyroxene, they form an angle of approximately 87° (Figure 2.2.14a). Also, as with amphibole, these cleavage planes often "block" one another. Longitudinal cuts through the crystal are, in comparison to amphibole, usually shorter and more compact. Thin exsolution lamellae are visible (Figure 2.2.14b) that at times reflect a weak zoning. Provided that pyroxene is light in color, it can be represented through light dotting. Oscillatory zoning in magmatic pyroxene (e.g., augite) is best represented through varyingly dense dotting (Figure 2.2.14c). However, the zoning is rarely as pronounced as it's sketched here.

2.5   Sketches for Fast Documentation

What has been said thus far informally results in a schema for quick sketching. The picture in plane-polarized light is always used as a template. Crossed polarizers are used to supplement. With their help, adjacent crystals of the same mineral can be better distinguished. "Fast documentation" means that the micrograph is drawn from scratch and only important parts are shown. What is important depends upon the purpose of the microscopy and documentation. Because a rock's mineral composition is widely defined and can deliver important information about its condition and history, the sketching of different minerals is in the foreground even during rapid sketching. Although sketches are usually labeled, the outline and internal fabric of individual crystals should be enough to distinguish minerals. It is usually sufficient to correctly render the few characteristic features covered in Chapter 2.4 (Figure 2.2). A precise and complete reproduction of the fabric is not necessary. Gaps are not only allowed, but advisable in order to keep the time requirement low. How well our brain can compensate gaps has been pointed out already. Even rough sketches should include a scale.

   The proportions between the individual parts in the fabric must remain reasonably intact. This is most easily achieved by mapping out major areas in the fabric (mineral layers, larger grains or grain groups, distinctive fractures, etc.) with light lines first and, if necessary, correcting them in the course of sketching. This means that the sketch is developed in several steps (Figure 2.3). In the final stages of the sketch, the minerals of higher refraction are represented in a heavy line weight and, possibly, with intensive dotting along the grain margins.

Figure 2.4 *Amphibolite-facies gneiss (Scheelite deposit Mittersill, Tauern Window, Eastern Alps, Austria); sample KR3768B; mineralogical composition: quartz (Qtz), amphibole (Am), biotite (Bt), chlorite (Chl), apatite (Ap) and scheelite (Tu); pencil drawing based on a microscope image in plane-polarized light; line weight 0.5 mm and locally lighter, due to obliquely set lead; original size of a single sketch*

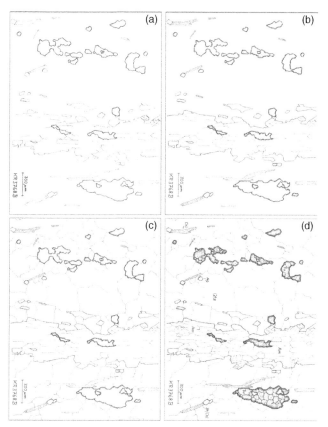

Figure 2.4 *(continued) ca. A4. The drawing develops successively. **(a)** Contouring of quartz areas in order to outline the layering and the grain distribution of other minerals; contours in light line weight for easier correction. **(b)** Intensification of contours and fillings according to the difference in refraction between quartz and other minerals. **(c)** Contouring within quartz areas with light line weight. **(d)** Complete filling of grains with high refraction, specifically of amphibole (with characteristic angle between cleavage planes) and scheelite; completion of grain boundaries in the amphibole layer (based on a microscope image with crossed-polarized light) and labeling. Grain boundaries in the scheelite areas illustrate the partitioning to larger grains with irregular shape and smaller isometric ones and thereby indicate the recrystallization of scheelite.*

Of course, even at the first stage of sketching, one must make sure that the different minerals retain their typical forms, for example, magmatic biotite with its thick, tabular shape and smooth crystal base or feldspar with its compact-elongate form (Figure 2.3a). In the next step, grain outlines are reinforced and internal fabrics drawn, for example, saussuritization and twins in plagioclase and cleavage planes in mica (Figure 2.3b). In the final step, fillings are completed, overlooked elements of the fabric are added, and the sketch is labeled (Figure 2.3c). The sketch must not have a fixed frame and single minerals can certainly "levitate." They just must fit into the fabric pattern.

If larger parts of the area to be drawn are occupied by a mineral of low refraction (e.g., quartz), it makes sense to first outline all other higher refractive areas in a thin line (Figure 2.4a), then to work out these areas through reinforcement of the outlines and partial filling, and only then to add the grain boundaries in the quartz area (Figure 2.4c). The completion of the internal fabric of the higher refractive minerals, as well as the labeling, is reserved for the last step (Figure 2.4d).

It is practical to group structural parts in a sketch from remote areas of a cut to avoid sketching unimportant intermediate regions (Figure 2.5). If grains are too similar with respect to line weights in the first phase of sketching, they should immediately be labeled (Figure 2.5a). Of course, the characteristic internal structure should, nevertheless, be represented in the next step of sketching in order to facilitate rapid recognition. But even here, it is sufficient to schematize the internal fabric (e.g. microcline twinning in K-Feldspar) (Figure 2.5b), as its formation and position in the crystal usually has no meaning.

Although the line weight cannot be easily varied with a pencil, at least the most important rock-forming minerals provide enough differences in their shape and internal fabric, so that (almost) no labeling is required. In mica-bearing rocks, mica is strikingly apparent in its tabular structure, its cleavage planes, and (in metamorphic rocks) its frayed crystal bases (Figures 2.6 and 2.7). The intrinsic color of biotite and chlorite, highlighted through dotting, offers a sufficient contrast to white mica. If minerals of similar refraction have a similar grain shape, like epidote—apatite, garnet—staurolite, or andalusite—feldspar, additional labeling is helpful.

Even with a black ballpoint pen, microfabric can quickly be put to paper (Figure 2.7). An advantage over the pencil lies in the fact that even thin lines and dotting can produce contrast. However, one needs experience and assurance in sketching, because corrections are either time-consuming or ugly.

In order to easily locate a small section in the total section area or to represent its position relative to larger structures, a small overview sketch of the total section area can be included (Figure 2.8). Even in this example, large areas of fabric are sketched lightly (Figure 2.8a) and then delineated from one another through line weight and internal fabric (Figure 2.8b). But the final version will not only be supplemented with labels, further internal fabric, and crystallographic directions,

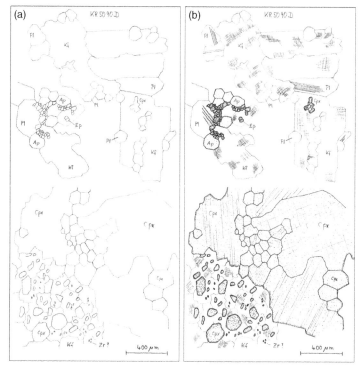

Figure 2.5 *Metasomatically overprinted conglomerate; Mary Kathleen Mine (Mt. Isa Inlier, Australia); sample KR5090D; mineralogical composition: K-feldspar (Kf), plagioclase (Pl), clinopyroxene (Cpx), epidote (Ep), apatite (Ap) and zircon (Zr); three pencil drawings based on microscope images in plane-polarized light; original size of sketch assembly ca. A4; development in two steps.* **(a)** *Contouring of all grains with the same line weight and labeling.* **(b)** *Intensification of grain outlines and characteristic internal fabrics: twin lamellae in plagioclase, diffuse gridiron (microcline) twinning in K-feldspar, and cleavage planes in clinopyroxene. These internal fabrics are schematized, that is, represent the real fabrics only roughly but characterize the minerals. The line weight of the boundaries between pyroxene and other minerals of lower refraction is increased, in contrast to the line weight of internal boundaries.*

but also by an enlargement of the section (Figure 2.8c). Such enlargements are always useful when major parts of the fabric are too small to be reasonably included to scale.

Fast, documentary drawing requires only partial accuracy and can be incomplete and limited to the bare essentials. Nevertheless, minimum standards should be respected with regard to the correct reproduction of minerals and fabrics. Fast, documentary drawing is an important (if not the most important) part of microscopy. It forces one to make exact observations and to think through what is being seen. Only with supplementary drawing can one soundly microscope. This means that one should always sketch and operate the microscope simultaneously.

Figure 2.6 *Amphibolite-facies andalusite schist (Sesia Zone, Western Alps, Val Loana, Northern Italy); sample KR1481; mineralogical composition: quartz, white mica, biotite, andalusite, staurolite and ore (opaque); pencil drawing based on microscope image in plane-polarized light; original size of drawing ca. A4. Biotite and white mica show outlines with similar line weights and cleavage planes typically pronounced close to the end faces of the thin tabular crystals. The brown, inherent color of biotite is represented by fine dotting, however, without considering the (meaningless) different intensity of color due to pleochroism. In comparison to mica, the sutured quartz grain boundaries, with their straight facets, are drawn with lighter line weight. The few subgrain boundaries are drawn as dotted lines. Based on the relatively high refraction of staurolite, the three small rounded grains (St) are silhouetted against the surrounding grain fabric by thick grain boundary lines and thick internal dotting.*

The drawing is dominated by the several mm large elongate andalusite grain (And). The evenly distributed fine pigment, characteristic of andalusite, is mimicked by light dotting. The drawing does not represent an exact copy of the thin-section image but is designed in such a way that the characteristics of the fabric can be recognized quickly and easily. Such a drawing provides a high information density and allows explicit conclusions about the rock's history of deformation and metamorphism. The slightly elongate ore and quartz inclusions in staurolite, partly linearly arranged, as well as the tiny white mica inclusions in coarse-grained quartz indicate an early foliation. These quartz grains probably overgrew the small plates of white mica and biotite after the formation of andalusite and during a continued or subsequent event of deformation under elevated temperature. Mica plates outside quartz coarsened considerably. The large andalusite grain is totally mantled by white mica. Size and random orientation of white mica indicate that it grew after the decline of deformation and at high temperatures.

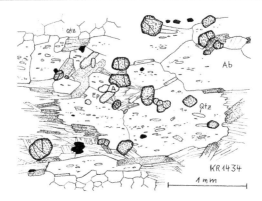

Figure 2.7 *Garnet-albite schist from the garnet zone of Barrow metamorphism (Dalradian of the Scottish Highlands; Power Station Ardlin, Loch Lomond, Scotland); sample KR1434; mineralogical composition: quartz (Qtz), albite (Ab), white mica, biotite, garnet, epidote (Ep), apatite (Ap) and ore (opaque); free-hand ballpoint pen drawing of thin-section image in plane-polarized light; variation of line weight through different pressure on the ballpoint pen; size of original drawing ca. A5. Garnet, epidote, and apatite grains serve as initial "support points." Subsequently, mica and albite grains are drawn and, finally, quartz and the inclusions in albite. In order to save time, the drawing is kept incomplete. Biotite and white mica show the same line weight but can be distinguished by internal dotting of biotite. Apatite (Ap) and epidote (Ep) differ from their albite-white mica surroundings by heavier line weight and from each other by different dotting that mimics the different refraction of the two minerals.*

The drawing does not represent an exact copy of the thin section image. However, attention was paid that diagnostically important characteristics are clearly visible. (i) Preferred orientations of mica and albite blasts, and partially banded orientation of white mica and quartz inclusions in albite, which all mark an early foliation; (ii) mica plates outside the albite blasts, which are oriented obliquely to the foliation, (iii) sutured grain boundaries and locally occurring subgrain boundaries (dotted lines), by which quartz can be distinguished from albite despite the similar line weights; (iv) the suturing of mica-albite boundaries, which points to growth of albite over mica.

Figure 2.8 *Breccia of the Zuccale Fault (Punta di Zuccale, Elba, Italy); sample ZV13a; mineralogical composition: quartz (Qtz) and calcite (Cc); pencil drawing of a thin-section image in plane-polarized light; size of original drawing A4. The drawing develops successively. (a) Contouring of the larger quartz and calcite*

(a)

(b)

(c)

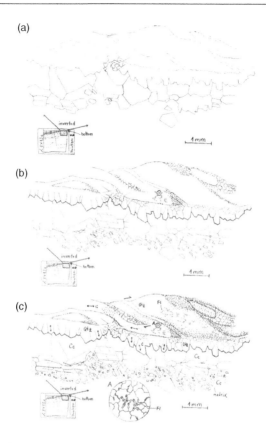

Figure 2.8 *(continued) areas, in order to outline the layering and record proportions and distribution of the crystals. In addition, (i) the position of the structure in thin section, (ii) the orientation of the thin section in relation to the sample and to the lineation ("striation"), and (iii) the orientation of the sample in relation to the fault ("bottom") are indicated. (b) Increasing of line weight of the palisade-quartz–calcite boundaries and drawing of fillings and small structures. (c) Drawing of (i) fine-grained areas, (ii) fluid inclusions (FI—only locally for saving time), and (iii) an enlarged section (A) for better visualization of fluid inclusions. Labeling and indication of dominant quartz-c orientations and the supposed shear sense.*

2.6 Development of Precise Thin-Section Drawings

What applies to a quick sketch, also applies to a precise thin-section drawing, regardless of whether it is made from a photo or screen template, with the help of a drawing tube, or freehand. The picture in plane-polarized light, with a shift to crossed polarizers as supplementation, always serves as the template for the drawing. With certain exceptions, the drawing is completed with an ink pen or a felt-tipped pen ("pigment liner"). However, the first version of the drawing should be done in pencil to allow for corrections. The size of the drawing depends on which details will be included. It is always advisable to start bigger and to downsize digitally later on, if needed. This way, the finest line can always be reduced another step.

As with a quick sketch, the grains of different minerals in the pencil version of a precise drawing should either be labeled or marked by typical internal fabrics—for example, through narrowly spaced cleavage planes in pyroxene and amphibole, curved fractures in garnet, or rough crosshatching in opaque grains, which is blacked out completely in the final drawing (Figures 2.9a and 2.10a). Such labels can be omitted if the minerals can be clearly distinguished from one another by their grain shapes, as in the case of mica and quartz, for example (Figure 2.11a). Nevertheless, even in this case, it makes sense to highlight the characteristics of minerals: in mica, the clearly visible cleavage planes at the base of the crystals; in quartz, the grain boundaries sutured in straight segments (Figure 2.11b); or, in pyroxene, the narrowly-spaced, parallel cleavage planes or fine exsolution lamellae (Figure 2.9b). Such strict parallel lines, as well as the thin, parallel twin lamellae in plagioclase, are best drawn with a ruler.

Figure 2.9 Metagabbro of the Finero Complex (Ivrea Zone, Southern Alps; Valle Cannobina, Northern Italy), deformed under amphibolite-facies conditions; sample KR740; mineralogical composition: garnet (Grt), clinopyroxene (Cpx), amphibole (Am), plagioclase (Pl), apatite (Ap), and ore (opaque); original size of the drawing ca. A4. *(a)* Pencil drawing of a thin-section image in plane-polarized light; all grain outlines and internal fabrics are drawn with equal line weight; minerals are labeled and marked by hatching and internal fabrics; ore grains are outlined in black. *(b)* Drawing finalized with ink pen. The high relief of garnet is represented by line weight 0.5 mm and dotting along the grain boundary; the high chagrin by thick internal dotting (line weight: 0.5 mm). The outlines of pyroxene and amphibole grains are drawn with line weight 0.35 mm. The closely spaced parallel lines in spotted areas represent fine exsolution lamellae. The brown, inherent color of the amphibole is silhouetted against the light green of pyroxene by dense dotting. This contrast clearly visualizes the amphibole-pyroxene intergrowth. Plagioclase is kept blank, in order to increase the contrast to the other minerals.

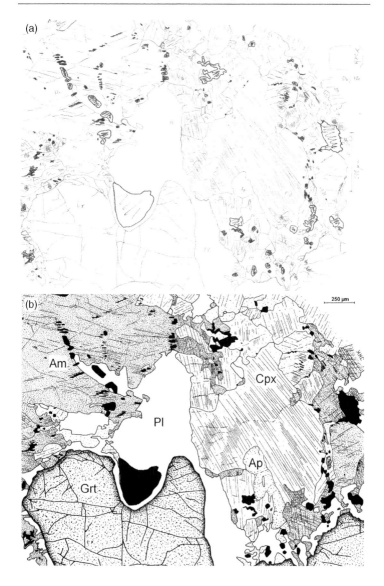

(a)

(b)

250 µm

Am

Cpx

Pl

Ap

Grt

The pencil sketch of the grain boundaries is traced with an ink pen, with different line weights for different minerals. It makes sense to sketch one mineral at a time rather than working on whole areas (e.g., from left to right). On the one hand, the pen doesn't constantly need to be changed, and, on the other hand, errors in line weights may be minimized in this manner. This is best done by starting with the highest refractive mineral (= heaviest line weight), sketching the outlines of all grains of this mineral, and then working on the other minerals in order of decreasing refraction. Those who would like (and have the time), can of course also scan the pencil drawing and re-work it with an image editing program. For complex

(a)

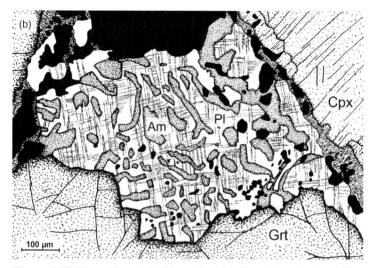

Figure 2.10 *Metagabbro of the Finero Complex (Ivrea Zone, Southern Alps; Valle Cannobina, northern Italy) formed under amphibolite-facies conditions; sample KR741-3; mineralogical composition: garnet (Grt), clinopyroxene (Cpx), amphibole (Am), plagioclase (Pl), and ore (opaque); drawing of a thin-section image in plane-polarized light, generated in A4 size with the aid of a drawing tube.* **(a)** *Pencil drawing of grain outlines and internal fabrics with equal line weight.* **(b)** *Ink pen drawing on tracing paper, based on (a). The contrasts in color and refraction of the different minerals are illustrated by different line weights and fillings: garnet 0.5 mm line weight, amphibole 0.35 mm, clinopyroxene 0.35 mm. Strong chagrin and high relief of garnet are represented by internal dotting and dotting along the grain boundaries (0.5 mm). Dense dotting (0.35 mm) and thick dotting along grain boundaries imitate the dark green inherent color of amphibole and the relief of the amphibole versus the surrounding plagioclase, respectively. Amphibole and plagioclase form coarse symplectite between garnet, clinopyroxene, and ore grains. Light-green clinopyroxene is optically separated from amphibole by light dotting (0.35 mm) and parallel cleavage planes. Although the twin lamellae of plagioclase are not visible in plane-polarized light, they are drawn as dotted lines in order to visualize the small distortions of crystal parts of plagioclase. The fine symplectite along the clinopyroxene margin, with small ore grains enclosed (black), is represented by approximately parallel lines (line weight: 0.25 mm).*

grain fabric, it makes sense to finish the black and white line drawing before drawing the internal structures, or simply to hint at the internal structures. Otherwise, outlines and internal structures can easily be confused in the maze of lines.

In detailed drawings of fabrics, internal structures should be highlighted and drawn precisely. Precision is important. The cleavage planes rarely intersect one another. Frequently, one cleavage plane terminates when meeting another one (Figures 2.2.11, 2.2.13, and 2.2.14). This hierarchy mirrors the different opening times of the individual cleavage planes. Grain-boundary segments in quartz (but also in other minerals: olivine, feldspar, calcite) should be drawn straight and not curved (Figures 2.2.1, 2.2.4, 2.2.8b, 2.2.9, 2.4, 2.6, 2.7, and 2.11), even when these segments, as a result of cut effects, appear curved under the microscope. Fractures in garnet and olivine, however, must be represented as curved and in hierarchical structure (Figures 2.2.5 and 2.2.8a). Cleavage planes in mica should predominantly appear on the (often craggy) crystal bases and less in the grain interior.

In general, the line weights of the internal structures should be somewhat less than those of the grain boundaries, because there is no difference in refraction on either side of a fracture, for example. However, this reduction in line weight only comes to fruition in higher refractive minerals. As soon as the fractures are open (as they often are, like grain and phase boundaries), they appear in plane-polarized light in the thin section as strong, dark lines. This is then also how they should be drawn.

In larger crystals, the inherent colors can only be imitated partially by a "compression" of the internal structures, which would be too great a deviation from the natural conditions. The inherent colors of biotite and amphibole, for example, are created through fine dotting (Figures 2.7 and 2.9b). The same goes for the

---

Figure 2.11 Mica schist of the Silbereck succession (eastern Tauern Window, Eastern Alps, Austria); sample KR2901B; mineralogical composition: quartz, white mica, epidote; drawing of a thin-section image generated in A4 size with the aid of a drawing tube. (**a**) Pencil drawing with equal line weight for all minerals. (**b**) Ink pen drawing of (a) on tracing paper; line weight of white mica 0.35 mm, quartz 0.25 mm, and epidote 0.35 mm; dotted roundish area in quartz: hole filled with epoxy resin. During revision, the cleavage planes at the ends of the white mica platelets are intensified in order to increase the contrast between white mica and the surrounding minerals. Already in the initial pencil drawing, the quartz grain boundaries are delineated as straight segments with sharp corners, characteristic of quartz, but not rounded as often shown in quick sketches. Further characteristics of quartz include: (i) accentuated 120° angles at triple junctions, often accomplished by short bends at the triple junction, and (ii) quartz grain boundaries that meet the flat faces of micas at right angles, again accomplished by short bends at the mica face.

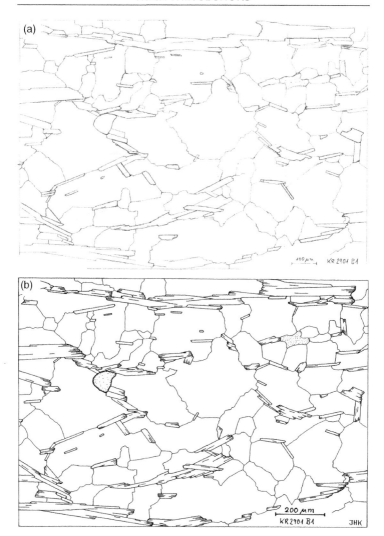

(a)

100μm   KR 2901 B1

(b)

200μm
KR 2901 B1   JHK

65

dark pleochroic halos in biotite or in hornblende, that are best represented through fine, dense dotting. Through dotting of different densities—not just small or large dots!—the differences in color between otherwise similar minerals, like amphibole and pyroxene, can be made clear (Figures 2.9b and 2.10b).

When colorless minerals with low refraction, especially quartz and feldspar, make up the "background" of the fabric, it makes sense to leave these areas empty in order to highlight more strongly the outlines of the remaining mineral grains (Figure 2.9b), unless, of course, the internal fabric conveys important information (Figure 2.10b). Even without grain boundaries, grain shapes can be represented with help of the internal fabric, for example, when fine fillings are available that mimic the grain shape (Figure 2.12).

Internal structures that are difficult or impossible to detect in plane-polarized light (especially subgrain boundaries, twin boundaries, deformation lamellae, narrow kink bands) are shown either with or without dotted lines. This dotting should always be done with small dots. To keep the drawings "optically easy," it is important to draw selectively, that is, no internal structures that are unimportant or deemed unimportant (e.g., grain alterations, saussuritization of plagioclase, pinitization of cordierite, serpentinization in olivine). If such structures are important, however, to mineral identification, they should be drawn. Subgrains typically occur in quartz (Figures 2.2.1, 2.6, and 2.7) but also in olivine (Figure 2.2.8b). Fine alterations to white mica along the grain boundary or fractures is of diagnostic importance in cordierite (Figure 2.2.12b). Exsolution lamellae and microcline twinning (difficult to draw!) in K-feldspar (Figure 2.2.3) and deformation twins

---

Figure 2.12 *Dark gabbro from Mailam/Pondicherry (southern India); sample KR5018; mineralogical composition: clinopyroxene, plagioclase, and ore.* **(a)** *Photomicrograph (plane-polarized light). The up to 1 mm large colorless plagioclase is tabular, free of cleavage planes and shows pigmented dark cores. The calcium-rich crystals (labradorite with An content of ca. 55 to 60%) form a grid with clinopyroxene filling. The clinopyroxene is anhedral. It is intergrown with massive ore (opaque) and segregates fine particles of ore.* **(b)** *Drawing of the photomicrograph (a); ink pen on tracing paper placed on the photomicrograph; original size ca. A4. The different pigment content of the plagioclase is highlighted by different dotting. Thus, the tabular shape of the crystals and their interlocking are visualized without drawing the grain boundaries. The outlines of the higher refractive clinopyroxenes are represented by dotting closely along the grain boundaries—not by a higher line weight. The visual contrast to plagioclase is increased by additional internal dotting, which also reflects the weak zoning, and by fractures (0.25 mm line weight, like the grain boundaries).*

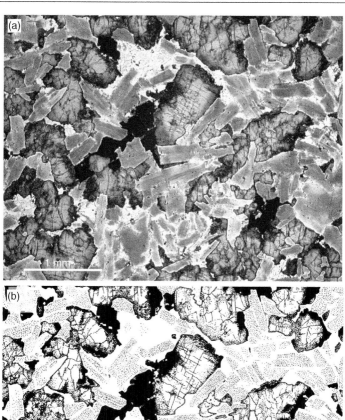

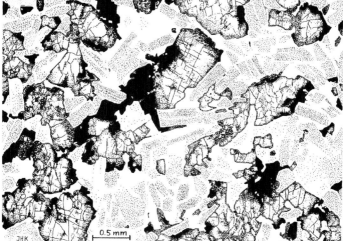

in plagioclase (Figure 2.2.2b) help to identify both minerals in thin section and keep them separate in thin-section drawings.

Drawing a highly structured fabric with many minerals that are similar in part is where it becomes difficult (Figure 2.13a). In this case, it is useful, (i) to increase the contrasts between the grains either by strengthening or weakening the internal fabric, for example, or by leaving low refractive minerals (usually quartz or feldspar) blank, (ii) to group areas of the same mineral grains together, that is, omitting grain boundaries or simply hinting at them, (iii) to visually highlight highly refractive grains with a thin, point and line-free halo (Figure 2.13b).

Figure 2.13 Kinzigite of the fossil Variscan continental lower crust (Serre/ Calabria, southern Italy); sample KR3279; mineralogical composition: garnet (Grt), quartz (Qtz), biotite (Bt), sillimanite (Si), cordierite (Crd), and ore (opaque). *(a)* Photomicrograph in plane-polarized light. *(b)* Ink pen drawing on tracing paper placed on a paper print of the photomicrograph; original size ca. A4. One purpose of the drawing is to show the distribution and random orientation of the sillimanite needles and the high amount of cordierite. Neither grain boundaries nor internal fabrics of quartz are displayed, and, therefore, quartz forms the white background of the drawing. This conforms to the relatively lowest refraction of quartz and, in addition, increases the total range of light-dark contrast in the drawing.

In the A4 original, the line weights of the different minerals are as follows: biotite 0.17 mm, cordierite 0.25 mm, sillimanite 0.35 mm, and garnet 0.5 mm. The contrast between biotite and cordierite is increased in two ways. First, the dense, even dotting makes the biotite appear relatively dark. Secondly, short and dense hatching along grain boundaries and fractures imitates the characteristic pinitization of cordierite and leads to a more pronounced edging of the grains. This, together with the characteristic pleochroic halos around zircon, allows us to identify cordierite even without labeling. The pleochroism-related variable inherent color of biotite is not considered, and the outlines of the crystals are not shown or only weakly delineated. Thus, unnecessary complexity of the drawing is avoided. The different orientations of the biotite crystals can be represented by different grain shapes. Identification of sillimanite is relatively straightforward, based on the thin columnar shape, the rhombic cross section, and the characteristic cleavage. In order to increase the contrast and visibility of the sillimanite needles, dotting in biotite close to sillimanite is omitted. If the shape of quartz grains is not significant, it is reasonable to leave the quartz regions completely blank. The other end of the scale is generally occupied by garnet that shows the highest refraction and strongest chagrin of all rock-forming minerals. The high refraction is represented by thick dotting along the grain margin and the high chagrin by intensive internal dotting.

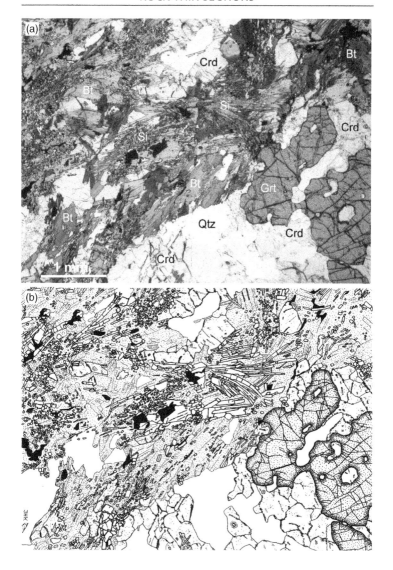

Especially in the comparison of photomicrograph, preliminary pencil sketch, and final ink drawing, ones sees the advantages of the ink drawing (Figure 2.14). In photomicrographs, visually similar minerals, like olivine and clinopyroxene, in this example, are difficult to keep apart, and even the identity of plagioclase can only be presumed due to the mineralogical composition of the rock; optically, it would not be recognized as such (Figure 2.14a). What is hard to make out in the preliminary pencil sketch (Figure 2.14b), gains visual prominence in the final ink drawing (Figure 2.14c). With meaningful internal fabric and clearly graded line weights, the various minerals are easily kept apart.

The way in which the internal fabric is drawn, depends on the size, or rather the resolution, of the drawing. In overview drawings of entire thin sections, the individual grains are usually relatively small and the internal structures not or hardly representable. In this case, a sparing, schematic filling of grains with internal structures is useful. These help to characterize individual minerals and separate them from one another. In small grains, they also serve to heighten contrast and to emphasize inherent color. This means that cleavage planes, for example, are narrowly spaced in pyroxene and relatively wide-spaced in feldspar to maintain the difference in brightness between the two minerals.

In such overview drawings, it is also useful to save time through omission and to make the drawing visually simpler. If the drawing is too full, contrasts disappear

---

Figure 2.14 *Gabbronorite from the Harz Mountains (Germany); sample M6; thin-section collection of TU Munich Geology; mineralogical composition: plagioclase (Pl), amphibole (Am), clinopyroxene (Cpx), olivine (Ol), apatite (Ap), and ore (opaque); long side of photo equivalent to 5.7 mm; original size of a single drawing ca. A4. (a) Photomicrograph in plane-polarized light. (b) Pencil drawing of grain boundaries in (a) on tracing paper placed over a print of the photomicrograph. Ore grains are marked by x. (c) Based on (a), the ink drawing is successively developed from minerals of high to minerals of low refraction. (i) Blackening of the ore grains; (ii) drawing of outlines of all mineral grains with appropriate line weights, from high to low: olivine 0.7 mm, pyroxene and amphibole 0.5 mm, plagioclase 0.25 mm; (iii) filling of the olivine grains (fractures and dotting along the grain boundaries and internally); (iv) filling of the clinopyroxene grains (cleavage planes, exsolutions [represented by short thick strokes], light internal dotting); (v) dense internal dotting of the amphibole grains, in order to generate the brightness contrast to pyroxene; (vi) characterization of plagioclase by twin lamellae (dotted lines with 0.25 mm weight), although the twins are only visible in crossed-polarized light but not in the photo (a). (d) Digital processing of the line drawing with four shades of gray.*

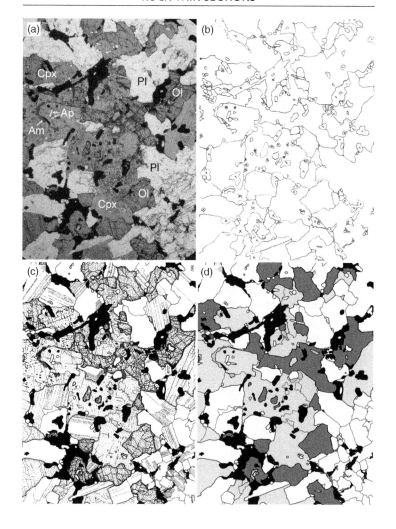

and detailed structures are difficult to detect. It is also a psychological problem, which primarily occurs through (but is not limited to) a lack of drawing practice. It is not easy to resist the urge to fill empty spaces in the drawing. If one notices that the drawing is too full, it is often too late. This can only be combatted with consistent practice or through digital correction of the image.

In overview drawings of rocks with high mica content, like schists, for example, micas are best drawn with thin, short lines (Figures 2.15 and 2.16). This way, foliations and folds are displayed prominently. The length of the lines allows for an estimate of the size of the mica plates. The distance between the lines gives an account of the mica content. Mica-poor (and mostly quartz-rich) layers are characterized by larger distances (Figure 2.15), whereas mica-rich layers or foliations, that develop from a folding, have smaller distances (Figure 2.16). In both cases, one creates a drawing whose impression is eerily similar to that of a photomicrograph. In compact quartz layers, it makes sense to only partially draw the grain boundaries to hint at the quartz fabric (Figure 2.16). But again, caution is needed because a drawing with too many lines is easily overloaded.

In overview drawings of quartz-feldspar-rocks, uniformly fine-grained areas are best represented through uniform dotting (Figure 2.17). Just as in thin section, these will then appear darker than the coarse-grained quartz and feldspar areas. Furthermore, fracture patterns are best only hinted at. In addition to the large amount of time required, an exact representation would fill the drawing too much and conceal important fabric characteristics. A completely blank depiction of the quartz "background" ensures a good breadth of contrast in the drawing. Whenever

Figure 2.15 Greenschist-facies metapelite of the Variscan basement (Argentiera, northwest Sardinia, Italy); sample SV180; mineralogical composition: quartz (Qtz), white mica (Wm) and ore (opaque); pencil drawing of a photomicrograph in plane-polarized light; original size of the drawing ca. A5; line weight 0.5 mm; thinner lines generated by obliquely oriented lead; SS = bedding; S1 = foliation of the first deformation event; S2 = foliation of the second foliation event (crenulation cleavage). White mica platelets are represented by strokes. The ore particles and masses are kept black. Quartz grain boundaries are not shown, in order to avoid strong filling and, consequently, illegibility of the drawing. The fine internal fabric of larger ore masses reflects the orientation of the foliation and is represented by the appropriate orientation of strokes (from lower-left to upper right).

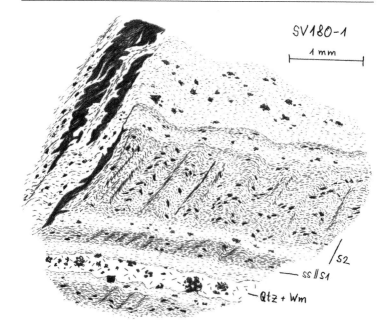

SV180-1

1 mm

/S2

—— SS ∥ S1

—— Qtz + Wm

Figure 2.15 *The drawing does not represent an exact copy of the thin section image but, due to time constraints, shows only the significant structures half-schematically. SS as well as the S1 orientation parallel to SS and the locally occurring crenulation cleavage S2 are well visible because only the mica platelets are presented and, therefore, the mica-rich layers are darker. These structures already reveal a lot of information. (i) The occurrence of S2 solely in mica-rich layers, (ii) the orientation of S2 at high angle to SS, and (iii) the orientation of small ore platelets parallel to S1 in mica-rich layers but their absence in quartz-rich layers are emphasized. Most probably, the quartz layer in the lower part of the drawing represents one of numerous quartz veins that typically form during the first deformation event parallel to S1. Even if the quartz grains of the vein are not drawn, their homogeneous size and generally isometric-polygonal shape can be inferred from the distribution and size of the mica platelets. The ore layers at the left margin of the figure are parallel to S2, but their substructures follow S1. This indicates that the ore was generated from a fluid during the S2 formation, that is, during the second deformation event. Further evidence of fluid-supported transport and formation of ore along grain boundaries is given by the partly reticular distribution of ore in the quartz vein and in the quartz-rich layers. A larger quadratic opaque grain at the margin of the quartz vein suggests pyrite as ore phase.*

the boundaries between mineral areas and a "main phase" are important, it is advisable to keep the main phase blank (Figure 2.18). This contributes to the highlighting of internal fabric along boundaries, like the symplectite rims of jadeite and garnet, in this example.

In overview drawings that cover a large part of the thin section, an even greater simplification and schematization is required. Fine and uniform structures in an overview sketch (e.g., the ground mass in a volcanic rock) are easily represented schematically by loose dotting (Figure 2.19), or, even more radically, by an open space. Drawing large-scale fillings only locally is time saving as well as visually effective. If the ground mass is well oriented, the preferred orientation of the small crystals can be indicated through short lines that are drawn only to the extent that the orientation is visible (Figure 2.20). Thus, the background of the sketch remains visually simple and the visibility of phenocrysts is preserved.

Figure 2.16 *Garnet-mica schist from the Dalradian of the Scottish Highlands (east of Ben Nevis); sample AH168B; mineralogical composition: garnet, white mica, quartz, epidote, and ore (opaque); ink pen drawing after a pencil drawing of a thin-section image in plane-polarized light, generated with the aid of a drawing tube; size of the original drawing ca. A4. The first foliation is built by white-mica platelets and tabular ore grains and is also preserved as relics of aligned and elongate quartz and ore grains in garnet blasts. The foliation is folded resulting in a weak crenulation cleavage. The high chagrin of garnet is represented by dotting with line weight 0.25 mm and the high relief by grain boundaries of line weight 0.5 mm with dotting along the boundaries (line weight: 0.35 mm).*

White mica is not drawn as single crystals but only indicated by thin strokes (0.25 mm line weight), which accentuate the orientation of the grains. Thus the matrix of the schist appears light without loss of details. Similarly, the boundaries of quartz in pressure shadows of garnet blasts are only locally adumbrated (line weight: 0.25 mm). Drawing these boundaries completely would make the quartz regions disproportionately dark. Epidote occurs as few elongate grains. The internal dotting (line weight: 0.25 mm) distinguishes the epidote from the less-refractive white mica and leads to visual separation from the white mica groundmass.

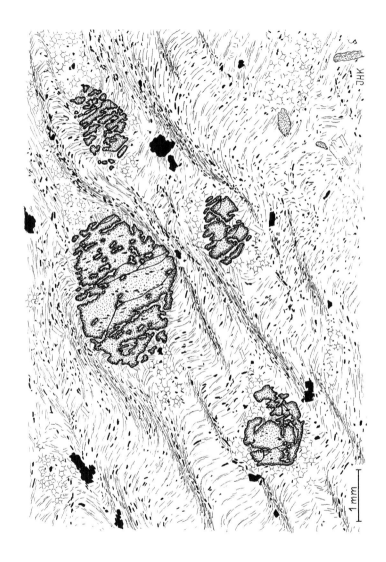

Figure 2.17 *Pegmatite from the fossil Variscan lower continental crust (Ivrea Zone, Alpe Scaredi, Val Loana, Southern Alps, Italy), strongly deformed under retrograde greenschist-facies conditions; sample KR2061. Ink pen drawing of a photomicrograph in plane-polarized light; original size ca. A4; line weight generally 0.35 mm, in plagioclase interior 0.25 mm. First of all, the drawing intends to show the difference between intensive brittle deformation in plagioclase (Pl) and crystal-plastic deformation in quartz (Qtz). The strong fragmentation of the large plagioclase grains predominantly parallel to cleavage planes is indicated by broken lines. Only local deformation twins of plagioclase are marked by thin parallel strokes. Grain boundaries in quartz layers are omitted, in order to increase the visual contrast to plagioclase. Quartz layers are locally lenticular and bent around plagioclase crystals, thus highlighting the strong crystal-plastic deformation of quartz. The fine-grained groundmass composed of quartz, plagioclase and white mica is represented by even dotting and, consequently, appears dark, corresponding to the thin-section image. The absence of grain boundaries in quartz regions intensifies the visual contrast and clarifies the coarse structure of the rock.*

Figure 2.18 *Jadeite quartzite (Shuanghe, Dabie Shan, China); sample RP01; mineralogical composition: quartz, jadeite, garnet, serpentinite along fractures in jadeite and garnet, albite, and symplectite at jadeite and garnet margins; drawing on tracing paper placed over the print of a thin-section scan in plane-polarized light; original size of the drawing ca. A3; felt-tipped pen with line weights 0.1 to 0.8 mm. Quartz grain boundaries are not shown, thus highlighting the geometry of the phase boundaries between quartz and the other minerals, specifically the cuspate structures along the foliation (ca. parallel to the long side of the drawing). Because pyroxene-typical cleavage planes are absent in the jadeite grains, the distinction from garnet is ensured by a specifically high visual contrast generated by dots and lines of high weight. Strokes of different length within the symplectite rims mark the different orientation of the symplectite columns relative to the thin section surface.*

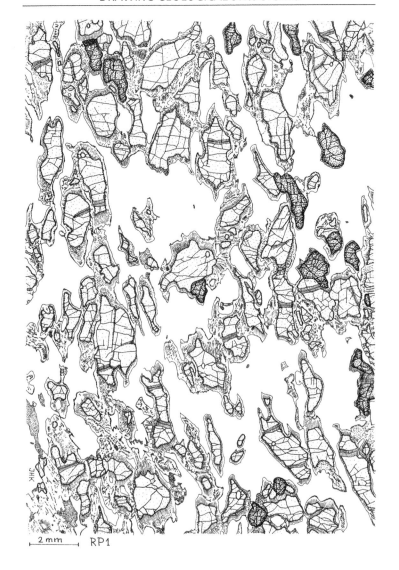

2 mm   RP1

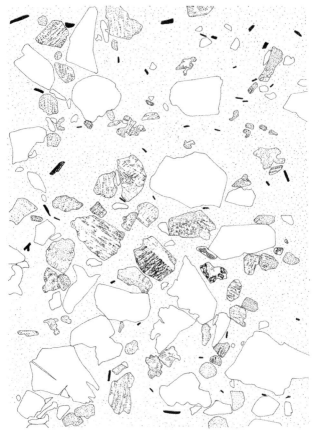

Figure 2.19 *Quartz porphyry (Bozen, South Tyrol, Italy); sample M64; thin-section collection of TU Munich Geology; mineralogical composition: (i) phenocrysts: ca. 0.5 to 1 mm large, subhedral to euhedral quartz, partially with resorption embayments; ca. 0.5 to 1 mm large K-feldspar and plagioclase; μm large thin tabular biotite; (ii) fine-grained groundmass: quartz, biotite, and feldspars. Quartz is kept completely blank; feldspars are filled with internal fabrics. For saving time and to increase the contrast to quartz, the groundmass is lightly and schematically dotted. Ink drawing of a microscope image (plane-polarized light) on tracing paper placed over a paper print of two merged photomicrographs; size of original drawing A3; line weight 0.25 mm.*

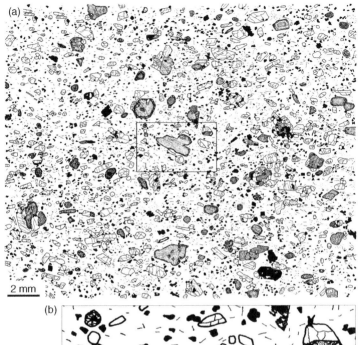

2 mm

←

Figure 2.20 *Tephrite (Kossal, Bohemia, Czech Republic); sample M45; thin-section collection of TU Munich Geology.* **(a)** *Ink pen drawing of six merged (ca. 30% overlap) photomicrographs (plane-polarized light) on tracing paper placed over a print of a photomicrograph; original size ca. A2; mineralogical composition: (i) phenocrysts: dark brown amphibole with reaction rims, subhedral light greenish-brownish clinopyroxene, and ore (opaque), (ii) fine-grained groundmass: dominantly thin-tabular plagioclase and ore. In the groundmass only larger plagioclase strips are delineated by short strokes, in order to keep the phenocrysts visible, in contrast to the photomicrograph. Similar to the amphibole and pyroxene phenocrysts, a locally weak "east-west" alignment of crystals is visible in the groundmass. In contrast to the groundmass dominated by plagioclase, amphibole and pyroxene phenocrysts exhibit a higher refraction, which is represented by higher weight of the crystal outlines (5 mm in the A2 original). The strong inherent color of the brown amphibole is illustrated by dense dotting which is absent in pyroxene with its clearly weaker inherent color. In addition, the dotting allows us to represent zoning visible in some of the phenocrysts, for example, in the upper right part of the image. Cleavage planes in amphibole and fractures in pyroxene require lower line weight (0.25 mm in the A2 original).* **(b)** *Detail from (a). Amphibole phenocryst with reaction rim. The uniform internal dotting only represents the inherent color, in contrast to the smaller pyroxenes without dotting. This blow-up also illustrates the local variation of the flow plane around the phenocrysts.*

Manual Drawing of Rock Thin Sections—Strategy:

- The microscope image (or a photo of it) in plane-polarized light serves as the template—the image through crossed polarizers serves as a supplement.
- Draw all grain and phase boundaries in pencil—no, or only hinted at, fillings.
- Set the line weight for the boundaries between different minerals (Figure 2.2): high difference in refraction = heavy line weight, but adjust line-weight schema for better contrast on a case by case basis; trace the pattern of boundaries with an ink pen.
- Be mindful that the difference in refraction at grain boundaries, the boundaries between grains of the same mineral, is minimal or nonexistent; if drawn at all, these boundaries should be drawn in a light line weight. This does not play a big role in minerals of low refraction (e.g., quartz, feldspar),

but does in minerals of high refraction (e.g., pyroxene, garnet). Pay attention to exceptions.

- Blacken opaque grains.
- If necessary, scan the drawing for digital reworking.
- Don't work on whole areas (e.g., from "left" to "right"), but rather one mineral at a time—from high to low refraction, that is, from heavy to light line weight.
- Schematize fine structures (e.g., ground mass in volcanic rock)—that is, dotting, drawing only locally, or leaving the whole area blank.
- Draw internal structures in large crystals:

  - Cleavage planes, fractures, and so on = solid lines, generally slightly less thick than the grain boundaries; pay attention to exceptions.
  - Twin and subgrain boundaries, and so on = only clearly visible with crossed polarizers which is why they are represented lightly (dotted, for example) and only if they are of diagnostic importance.
  - Highlight and precisely sketch, with a ruler if needed, mineral-typical structures (e.g., cleavage planes in amphibole).
  - Draw selectively—no unimportant structures (e.g., saussuritization in plagioclase—but be aware of exceptions).
  - Represent inherent color (e.g., for biotite, amphibole) through dotting of various densities that make the grain appear appropriately dark.
  - Represent chagrin through coarse internal dotting (or fine circles) and relief through dotting around the edges.
  - These steps generally apply to digital drawing as well.

2.7   Digital Reworking of Manual Thin-Section Drawings

If a thin-section drawing serves not only the purpose of rapid documentation or graphic observation, but rather the presentation in a publication or on a poster or for further investigations, a digital reworking presents itself. This can be carried out though simple image editing programs that are also available as "open source software."

Since it is already advisable to archive digital drawings (in sufficient resolution, i.e., with more than 300 dpi), the template for digital corrections and additions already exists. The standard steps of reworking include contrast enhancement and "sharpening" of the lines through "erosion" or "masking" functions. Furthermore, inaccurate lines can easily be eliminated or gaps supplemented. Labeling with printed characters or with lines, arrows, and other symbols, as demonstrated in many of the figures in this chapter, increases the readability and meaningfulness of a drawing.

82

If primarily the shape and distribution, rather than the internal structures, of crystals play a role, digital filling is a powerful tool. It helps to distinguish minerals of similar refraction and similar internal fabric from one another and to highlight small mineral inclusions (Figure 2.14d). In complex distribution patterns, digital fillings contribute to better visibility of the individual phases (Figures 2.21c,d and 2.22c, d). Usually filling is done in different shades of gray, because they are neutral and cannot mimic any internal fabrics. Here it makes sense to model the light-dark differences of the original manual sketch. Minerals with inherent color, dense internal fabric, or high refraction, thus, high relief and high chagrin, are given the darkest shades of gray (Figure 2.14d). It can certainly be irritating if this rule is violated (Figure 2.21d). Generally, however, the distribution and proportion of just a few phases can be made visually apparent this way (Figure 2.23).

2.8 Digital Drawing

Fully digital drawing has the advantage that (i) the drawing can be created directly from a digital photomicrograph without taking detours on paper, (ii) one can zoom into the photo while drawing, (iii) the line weights can easily (even retrospectively) be changed and (iv) one can work with (often more meaningful) color photos. Compared to the manual drawing of a thin section directly under the microscope, these advantages diminish in most cases. Adding natural boundaries with a mouse or "touch pad," is significantly slower and more strenuous. All too easily the lines appear angular and unnatural. Tracing simple boundaries in high resolution and then zooming out, leads to quite satisfactory results (Figure 2.24); correspondingly, however, this costs a lot of time for larger drawings. Digital drawing is almost completely useless when it comes to the creation of internal fabrics. The depiction of simple fracture patterns may still be satisfactory (Figure 2.24b). Impossible to represent digitally is a fine and irregular structure like that of symplectite, for example (Figure 2.24a). This leaves only the different areas to be filled with patterns and different shades of gray (Figure 2.24c). The result, however, is suitable only for a rough distinction between areas and different phases.

It is not recommended to use patterns (circles, lines, stripes, etc.) as filling. This always violates the appearance of the natural internal fabric, complicates the drawing, and confuses the viewer, rather than contributing to any kind of clarification.

Digital drawing if useful whenever few phases need to be delineated from one another, like the representation of the distribution and orientation of phenocrysts in a fine-grained ground mass, for example. In simple grain-boundary and fracture patterns, these can be traced in different line weights (or in different colors) in the digital photomicrograph without taking too much time (Figure 2.23a). Detached from the photo, they give a clear pattern from which preferred orientations and distributions can easily be seen and which can serve as a basis for fillings (Figure 2.23b) or further investigations (image analysis, quantification by means of fractal geometry).

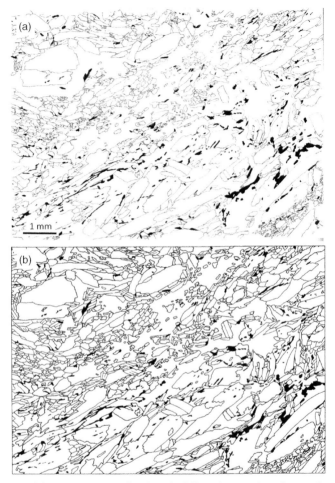

Figure 2.21 *Kyanite-mica schist from the Silbereck succession of metasedimentary rocks (eastern Tauern Window, Eastern Alps, Austria); sample KR2928C. (**a**) Thin-section drawing generated with the aid of a drawing tube (plane-polarized light); pencil on paper, original size A4; in the first step of drawing, kyanite was outlined and ore grains filled black. (**b**) Kyanite and white mica grains are outlined with ink pen, with higher line weight for kyanite. The quartz groundmass is kept blank.*

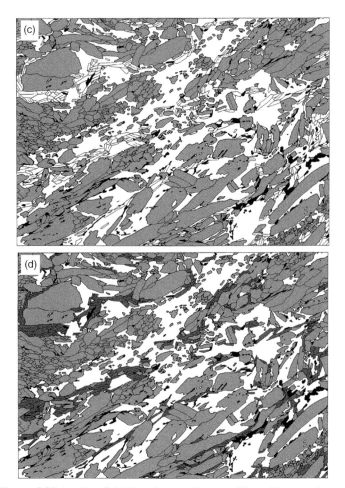

Figure 2.21 *(continued)* **(c)** *Digital processing (gray filling of kyanite) of a scan of drawing (b); gray level H,S,V,R,G,B = 0,0,82,210,210,210.* **(d)** *A darker gray filling of white mica is added to (c); H,S,V,R,G,B = 0,0,56,143,143,143.*

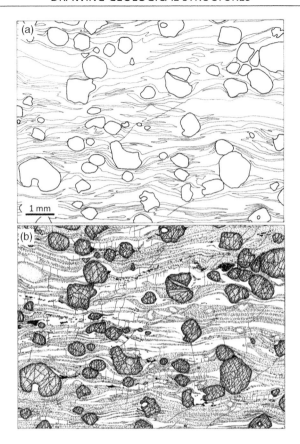

Figure 2.22 *High-temperature mylonite of a stronalite from the Ivrea Zone (Alpe Lut, east of Colloro, Val d'Ossola, Southern Alps, Italy); sample KR2980B; mineralogical composition: garnet, K-feldspar, and quartz. (a) Ink pen drawing from a thin section in plane-polarized light, based on a pencil drawing generated with the aid of a drawing tube; size of original image ca. A4; only outlines of garnet, and quartz and feldspar layers are presented together with few fractures; line weight for garnet 0.5 mm and for quartz and feldspar 0.25 mm. (b) Elaborated pencil drawing with fracture patterns (line weight: 0.35 mm) and dotting of the interior and along boundaries of garnet crystals, and with dotting of the finely*

Figure 2.22 *(continued) feldspar layers (line weight: 0.25 mm). Relics of coarser grains are accentuated by light dotting. The boundaries between quartz and feldspar layers are delineated as strongly dotted but discontinuous lines corresponding to the diffuse transition visible in the thin-section image. Grain boundaries of recrystallized grains in the quartz layers are omitted, in order to increase the contrast to feldspar layers. Additionally, cross fractures (line weight: 0.25 mm) characterize the quartz layers. Small roundish epidote grains (black) are supplemented, which are linearly arranged in quartz layers. (c) Digital processing of drawing (a) accentuates the garnet distribution, based on gray filling of the blasts. (d) The additional gray filling of K-feldspar layers clarifies the high K-feldspar content of the mylonite.*

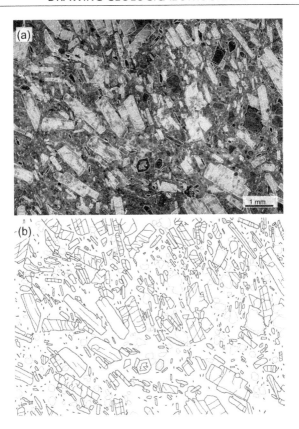

Figure 2.23 Aegirine-nosean phonolite (Eskişehir, Anatolia/Turkey); sample M102; thin-section collection of TU Munich Geology; mineralogical composition: (i) phenocrysts: tabular plagioclase, green aegirine with zoning, and strongly altered nosean that is only weakly silhouetted against the groundmass as diffuse light spots; (ii) fine-grained groundmass: thin-tabular plagioclase and pigment; size of original photomicrograph ca. A3, generated by two merged A2 images. **(a)** Photomicrograph (plane-polarized light) with white lines retracing (i) outlines and zoning of aegirine (line weight: 4 and 2 pixels), (ii) outlines and fractures of plagioclase (line weight: 4 and 3 pixels), (iii) outlines of nosean (line weight: 2 pixels), and (iv) the orientation of thin plagioclase platelets (short strokes; line weight: 5 pixels) generated by an image processing program. **(b)** Image (a)

Figure 2.23 (continued) *without photomicrograph as background. White lines are changed to black and gray. The groundmass is evenly displayed in light gray, in order to highlight the phenocrysts and their flow pattern (H,S,V,R,G,B = 0,0,95,243,243,243).*

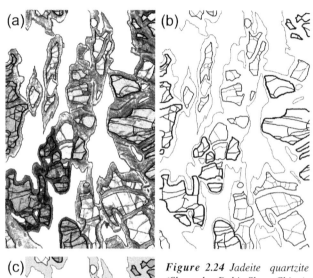

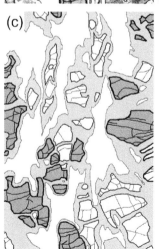

Figure 2.24 Jadeite quartzite *(Shuanghe, Dabie Shan, China); sample RP01; digital revision of a thin-section drawing with a standard image processing program; detail from Figure 2.18.* *(a)* Boundaries of albite and symplectite versus quartz (1 pixel line weight), outlines of jadeite (4 pixels) and garnet (8 pixels), and fractures in jadeite (2 pixels) and garnet (6 pixels) are retraced. *(b)* Pattern of grain outlines separated from the thin-section scan. *(c)* Grains of the different minerals are filled with different shades of gray.

Summary

- In contrast to a photo, important things can be highlighted in a thin-section drawing and unimportant ones omitted. Omission is always a good option while drawing. Drawings should stay "visually simple."
- Drawings are "cleaned-up" photographs—without cut effects, alterations, and preparation artifacts.
- Different minerals are represented primarily by different line weights, dotting, and internal structures.
- A high refraction requires a heavy line weight, possibly with additional marginal dotting. Inherent color can be represented by extensive, fine dotting.
- Precise drawings for publications are developed from pencil templates to ink drawings, not by area but from mineral to mineral starting with the highest refractive index—that is, the heaviest line weight.
- Thin-section sketches, created manually directly from the microscope, are a fast and effective form of documentation.
- A digital reworking can improve and complement the manual thin-section drawing. A completely digitally created drawing, however, is not a full substitute for a manual drawing.
- Drawing forces us to look at structures in thin section long and thoroughly ("guided seeing") and helps us identify and understand the fabric. This means: drawing should always accompany microscopy. Only with simultaneous drawing can microscopy be successful.

Exercise 2.1

Exercise 2.1 *Photomicrograph of a schist from the basement of the eastern Tauern Window (Eastern Alps, Malta Valley, Austria); sample KR3006A; plane-polarized light; short side of photo ca. 1.5 mm. Biotite layers (dark) form a foliation and are tightly folded. The light regions are predominantly composed of quartz. Place a tracing paper on the photo and generate (a) an ink pen drawing that shows the folded layers along general lines, (b), based on this draft, a more precise drawing with details of the biotite layers and quartz regions, and (c) quickly and "free-handedly" a pencil drawing of approximately the upper half of the photo, which shows the characteristics of the fabric without being precise.*

Exercise 2.2

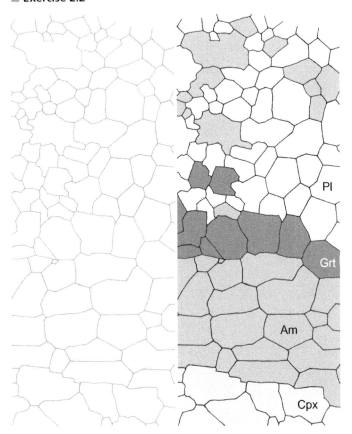

Exercise 2.2 *Thin-section drawing of a granulite-facies metanorite (Grenville Orogen, Canada); sample KR3768B; short side of image ca. 1.5 mm; section across a reaction zone between plagioclase (Pl) and clinopyroxene (Cpx), which is composed of garnet (Grt) and amphibole (Am). Fill the grains with typical internal fabrics, and bring the grain boundaries to their characteristic line weight.*

References

Bard, J.P. (1986). Microtextures of igneous and metamorphic rocks, D. Reidel, 269 pp.

Droop, G.T.R. (1981). Alpine metamorphism of pelitic schists in the south-east Tauern Window, Austria, Schweizerische Mineralogische und Petrographische Mitteilungen 61, 237-273.

Fueten, F. and Mason, J. (2007). An artificial neural net assisted approach to editing edges in petrographic images collected with the rotating polarizer stage. *Computers & Geosciences 33*, 1176-1188.

Hatch, F.H., Wells, A.K. and Wells, M.K. (1972). Petrology of the igneous rocks, 13th edition, Thomas Murby & Co, 551 pp.

Hoffman, D.D. (1998). Visual Intelligence: how we create what we see, W.W.Norton & Co., New York.

Kruhl, J.H. (1996). Prism- and basis-parallel subgrain boundaries in quartz: a micro-structural geothermobarometer. *Journal of Metamorphic Geology 14*, 581-589.

Kruhl, J.H. and Peternell, M. (2002). The equilibration of high-angle grain boundaries in dynamically recrystallized quartz: the effect of crystallography and temperature. *Journal of Structural Geology 24*, 1125-1137.

MacKenzie, W.S. and Adams, A.E. (1994). A Colour Atlas of Rocks and Minerals in Thin Section, Manson Publ., 192 pp.

Mason, R. (1978). Petrology of the metamorphic rocks, Allen & Unwin, 254 pp.

Moorhouse, W.W. (1959). The study of rocks in thin section, Harper & Row, 514 pp.

Okamoto, N. (1996). Japanese ink painting: the art of Sumi-E, Sterling Publishing Co., Inc.

Passchier, C.W. and Trouw, R.A.J. (2005). Microtectonics, 2nd edition, Springer, 360 pp.

Perkins, D. and Henke, K.R. (2003). Minerals in Thin Section, 2nd Edition, Prentice Hall, 176 pp.

Raith, M.M., Raase, P. and Reinhardt, J. (2012). Guide to Thin Section Microscopy, ISBN 978-3-00-037671-9 (PDF), 127 pp.

Richey, J.E. and Thomas, H.H. (1930). The geology of Ardnamurchan, north-west Mull and Coll, Memoir for geological sheet 51, part 52 (Scotland), British Geological Survey, 393 pp.

Tröger, W.E., Bambauer, H.U., Taborszky, F. and Trochim, H.D. (1979). Optical Determination of Rock-Forming Minerals, Part 1: Determinative Tables, Schweizerbart, 188 pp.

Vernon, R.H. (1986). K-feldspar megacrysts in granites – phenocrysts, not porphyroblasts. *Earth-Science Reviews 23*, 1–63.

Vernon, R.H. (2004). A practical guide to rock microstructure. Cambridge University Press, 594 pp.

Voll, G. (1960). New work on petrofabrics. *Liverpool and Manchester Geological Journal 2*, 503-567.

Voll, G. (1976a). Recrystallization of quartz, biotite and feldspars from Erstfeld to the Leventina Nappe, Swiss Alps, and its geological significance. *Schweizerische mineralogische und petrographische Mitteilungen 56*, 641-647.

3

SPECIMEN SECTIONS

Even drawing specimen sections helps document geological structures. The structures or the specimen sections are not simply depicted as accurately as possible, but rather, only the geologically relevant information is included. Therefore, one cannot simply "draw away," but must first decide what is technically important. Before drawing comes observation and, hand in hand with observation, goes interpretation. What is geologically relevant? Which minerals, structures, or fabrics are important for a particular question and must be represented? What is geologically insignificant and can be omitted? The observation, the professional interpretation, and the decision what to draw and what not to draw should come first and are more important than the drawing itself.

Drawings should document. This is only possible with schematization. It is important that this schematization follows clearly defined rules. Only if there is no arbitrariness and schemata are comprehensible, can structures in different drawings be compared. Geological drawing, therefore, must satisfy the claim to represent structures as accurately as possible but also schematically. This requires the symbolization of drawings, as discussed in the introduction, in which structures are not represented how they are seen, but in their "symbolism." The symbol must be complex enough to contain all the important information, but simple enough to be recognized quickly and easily.

3.1 The Geological Message of a Drawing

Even if geological structures aren't meant to be drawn realistically, the geological message of a structure must be conveyed correctly. This message may not be falsified by the inevitable simplifications and omissions of a drawing. What does and doesn't matter to the geological message should be decided upon, if possible, before starting the drawing.

It also follows that the "filling" of structures—for example, the filling of folded layers in a drawing—must be consistent with the form. Fillings that are modeled on the usual rock signatures in maps are useless for geological drawings. At best, they do not provide any information; at worst, they provide information that is at odds with the rest of the drawing and geologically incorrect. Limestone is usually symbolized through "masonry." This may be sufficient for area filling in a geological map (Figure 3.1a). When filling a limestone layer in a geological drawing, the

Drawing Geological Structures, First Edition. Jörn H. Kruhl.
© 2017 John Wiley & Sons Ltd. Published 2017 by John Wiley & Sons Ltd.

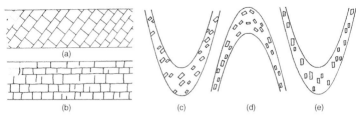

Figure 3.1 *Schematic fillings of rock structures. **(a)** Regular "brick" filling as characterization of limestone, with the joints oblique to the layer boundary; hardly acceptable in a geological map and definitely not in a geological drawing. **(b)** Filling of a limestone layer based on slightly irregular "bricks" with joints oriented parallel to the layer. **(c)** Folded layer of a porphyritic granite with feldspar phenocrysts whose flat faces are oriented obliquely to the layer and to the axial plane of the fold; already unacceptable in a geological map. **(d)** Flat faces of phenocrysts aligned around a folded layer. **(e)** Alignment of phenocrysts to a foliation parallel to the axial-plane of the fold.*

"wall" should always be parallel to the layer, like the individual layers in nature that are also parallel to the boundary of the entire limestone layer. The fractures should be sketched perpendicular to these layers and partially incomplete, as they are usually not continuously visible in natural limestone layers either (Figure 3.1b).

In a folded layer of porphyritic granite, the preferredly oriented feldspar phenocrysts should never be sketched lying at an angle to the fold (Figure 3.1c), but rather matching the folded schistosity (Figure 3.1d) or a schistosity corresponding to the fold (Figure 3.1e). Apart from extreme exceptions, this is the only way it looks in nature. But such exceptions must always be explained in the context of the entire drawing. What is disturbing in Figures 3.1a and 3.1c is mainly that the orientation (anisotropy) of the filling ("masonry brick," phenocryst orientation) does not coincide with the shape of the structure. Many geological structures are anisotropic and their fillings should be modeled on this anisotropy. This not only benefits the aesthetic feeling of the drawing, but also its informational content and significance.

Simple internal fabrics can already convey a wealth of geological information. Even the relationship between folding and schistosity can reflect different chronologies of deformation processes (Figure 3.2). If a layering-parallel schistosity is bent around a fold, it was already in existence before the fold (Figure 3.2a). If foliation planes are arranged symmetrically to the axial plane of fold, they were formed during folding (Figure 3.2b). Furthermore, foliation planes in fan or pile positions indicate different strengths of rock layers. If mica grows completely irregularly in a folded rock layer, it shows that its formation occurred after folding (Figure 3.2c). These internal structures can be depicted using just a few strokes.

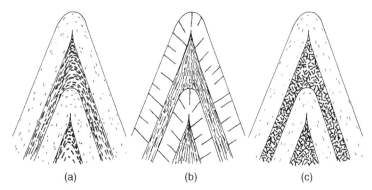

(a)   (b)   (c)

Figure 3.2 *Alternating sequence of folded quartzite-schist layers, with three possible representations of internal structures. Original drawing 15×7 cm; felt-tipped pen with 0.1 mm line weight.* ***(a)*** *The two rocks are foliated prior to folding. Accordingly, the schistosity planes are folded. In the schist, the schistosity is formed by short mica films and, in the quartzite, by oriented quartz grains. The schistosity is sufficiently depicted by a few strokes, which also accentuate the contrast in lightness between the two rocks.* ***(b)*** *The two rocks were foliated during the early phase of folding. In the relatively stiff quartzite layers, the schistosity developed as fractures in fan position. It forms planes in pile position in the schist.* ***(c)*** *The two rocks were foliated prior to folding. In the quartzite, the schistosity is folded. In the schist, new mica is formed after folding by a subsequent metamorphic reaction resulting in a random orientation of the mica platelets.*

They don't need to be sketched precisely or completely. They also need not be scaled in accordance with the size of the fold. The size of the fold does not matter in the context of the aforementioned geological messages. Nor is it important that the foliation planes take on an exact fan or pile position, or run exactly parallel to each other. Their spatial relationship with respect to the folded layers must, however, be represented correctly, for it includes significant statements about the sequence of deformation processes.

3.2 Schematic Representation of Minerals

Chapter 2.5 showed the importance of representing minerals in their characteristic forms, as part of documenting fabrics in thin section, to heighten their recognition value. The same applies when drawing rocks and their structures.

(Almost) every mineral has a typical form that should appear in the drawing if the scale permits. These forms, even if simplified as necessary, correspond to the crystal outline, which we know from the thin section drawing (Figure 2.1). These may be supplemented with the mineral typical line weights and internal structures, which are also used in thin-section drawings (Figure 2.2). Again, it depends on the scale. Drawings of specimen sections or parts of an outcrop face can contain details that resemble those in a thin-section drawing (Figure 3.3). But one must not stick slavishly to a particular schema. Although the outlines of euhedral feldspars in a porphyritic dyke are typical and need no further emphasis to make the mineral recognizable (Figure 3.3b), this is not always the case for other minerals. The fine grain fabric of deformed and recrystallized quartz-crystals can be modeled through schematic dotting and contrasted with feldspar, if the latter is kept blank. A similar contrast can be generated vice versa, that is, by keeping the quartz blank and filling the feldspar (Figure 3.3c). As in the present example, since the strong orientation of the quartz lenses reflects the schistosity of the rock well, it is not necessary to depict this in another way. If, however, the irregularity of the flow fabric is considered important, it can be shown with short lines that model the orientation of small biotite platelets (Figure 3.3c). Which part of the rock and mineral fabric is shown and in which way brightness contrasts are generated depends on what is deemed important. Only that should be drawn. Omission is also advisable when drawing rock sections.

Figure 3.3 *Photo of a rock cut and two schematic drawings. Both drawings highlight different aspects of the rock fabric. Scale always as in figure (c). Quartz-porphyritic dacite; Sesia Zone, Val Loana (Western Alps, northern Italy). (a) Sample KR2261; photo of polished section perpendicular to the foliation. The euhedral, white feldspar phenocrysts clearly stand out from the fine-grained dark ground mass. In contrast, the deformed lenticular and transparent-gray, mostly elongate quartz grains are barely visible. (b) Line drawing of feldspar phenocrysts and quartz lenses. This drawing best shows sizes and distribution of quartz and feldspar. Feldspars and groundmass are kept blank. The internal dotting of the quartz lenses mimics the complete, fine-grained recrystallization. The shape contrast between both minerals indicates that deformation took place under conditions of crystal plasticity of quartz and brittle behavior of feldspar. (c) Line drawing with emphasis on internal fabrics of feldspar and the groundmass. The quartz lenses are kept blank. Compared to (b), the contrast between quartz and feldspar is generated by blank quartz and filled feldspar. The indistinguishable flow fabric and schistosity in the groundmass are formed by large, magmatic biotite platelets up to 100 μm in size and represented by short strokes. While shape and orientation of the quartz lenses reflect roughly homogeneous planar flattening of the rock, the orientations of biotite in the groundmass indicate a more complex flow and deformation pattern on a smaller scale. Felt-tipped pen with line weights of 0.1 to 0.2 mm; original size of both drawings ca. A4.*

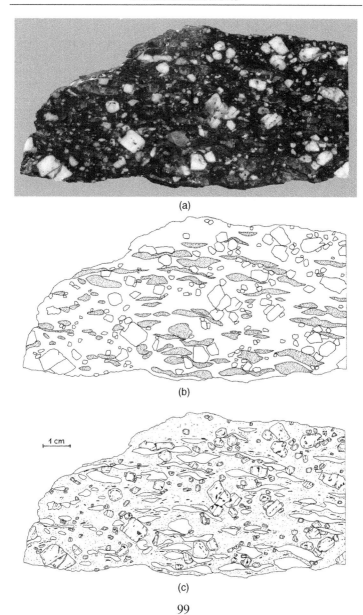

(a)

(b)

1 cm

(c)

3.3 Schematizing Rocks and their Structures

In all drawings that are done in the field or from samples and sample cuts, we draw rock structures. It is therefore important, that these structures be represented so characteristically that they enable identification of the rock without detailed labeling. In rocks that closely resemble each other this can be difficult, but for most it works quite well if a few rules are followed.

Moreover, it is not just about the identification of rocks but primarily about the representation of their fabrics. A rock's history (deposition, crystallization, deformation, metamorphism), or at least part of it, can often be derived from the rock's fabric. The emphasis in such a drawing should not be on the number of foliation planes per decimeter or the number of mineral grains per unit area or the exact size of folds or the thickness of layers. The point is rather to convey a geological message with such a drawing. From this follows that these drawings should neither be a one-to-one copy of nature nor drawn exactly to scale.

What is meant by this, will be explained by means of referencing drawings of some common rocks (Figure 3.4). Sketching sedimentary rocks is a thankless task in a way, because their structures—beyond just the layering—are not always clearly developed and the grain fabric is often not readily visible. One can, however, differentiate "sandy" sedimentary rocks with differing grain sizes from one another using dottings of different line weights or fine circles (Figure 3.4a). Even cross bedding, for example, and clay interlayers can be represented without problem. The more internal fabric a layer has, the easier it can be represented. Fragments in volcanic rocks are an example of this (Figure 3.4b). They characterize a rock not only by size and shape but also by orientation and internal fabric. This means: the fragments must be large enough to be drawn—oversized, if necessary, to make way for internal fabric or precise contours.

Limestone can only be graphically characterized by layer thickness and the pattern of (cross)fractures (Figure 3.4c). Where necessary, fossil-symbols can constitute a useful supplement. The characterization options in igneous rocks are also limited. In granitoid rocks one should limit oneself to representing biotite platelets and, possibly, the cleavage planes of coarse-grained feldspar (Figure 3.4d). Grain size, orientation, and homogeneity or inhomogeneity can be represented with biotite strokes alone. If larger euhedral feldspars occur in granitoids, a flow pattern can be sketched with them (Figure 3.4e). But watch out! Especially here, the crystal distribution should be "fractal" as it is in nature also. A uniform crystal distribution looks unnatural (see also Chapter 1.7). Gabbro and diorite frequently (but not always) have a random fabric. This is most easily illustrated by short, thick or thin lines that model the dark appearance of the two rocks (Figure 3.4f).

With fine-grained, homogeneous metamorphic rocks such as quartzite, there exists a problem similar to that of sedimentary rocks: often there is little to no distinctive fabric. Again, one makes do with different dashes or dottings with which layering, different mica content, and any existing schistosity can be represented (Figure 3.4g). It is important that these fillings not be too close, so that the generally lighter appearance of quartzite, as compared to other rocks, comes to light.

There are various formats for the representation of foliated metamorphic rocks, through which the intensity of deformation and degree of metamorphism, or different deformation phases, can be showcased. This is how the schistosity of augengneiss deformed under greenschist facies conditions, for example, can be represented using short, almost parallel-oriented strokes and rectangles. The lines represent the arrangement of biotite flakes and the rectangles the arrangement of feldspars (Figure 3.4h). The fractures in the feldspars and the rotated parts of the crystals are important components of the drawing. They point to greenschist facies conditions or lower temperatures during deformation. At higher temperatures, feldspar behaves plastically and lenses of recrystallized feldspar grains softly enveloped in foliation planes are formed (Figure 3.4i). The different stretching of the lenses reflects the different intensities of deformation. The inclusions typical of feldspar blasts can be well characterized through dotting (Figure 3.4k). Garnet porphyroblasts are drawn with a greater line weight compared to feldspars and with the curved fractures typical of garnet. The line weight for the outline and the fractures stays the same, because, in contrast to the thin-section image, the light refraction in a sample drawing doesn't matter.

Clearly developed foliation structures in rocks, like crenulation cleavage mostly composed of continuous mica films, for example, are most easily drawn with solid lines (Figure 3.4l), or with dashed lines, if the surfaces are less dominant or the drawing should be kept simple. Quartz veins, preferably those that form parallel to the first foliation during the first deformation of mica- and quartz-bearing rocks, are important for the deformation analysis of such rocks and should always be drawn concisely—especially with their typical lenses and tattered isoclinal fold crests (Figure 3.4m). In a drawing with many foliation planes, the quartz lenses and layers are best kept blank, to retain their contrast to the rest of the drawing. Only cross fractures can be supplemented. They are typical of quartz layers and contribute to their easy identification.

In metapsammopelites—metamorphic rocks consisting of layers of different quartz and mica content—one can use simple strokes to work out the different deformation behavior of the layers, e.g. through various forms of boudinage (Figure 3.5). In the simplest case, the foliation planes are represented using solid lines. The closer they are to one another, the more strongly the rock is foliated. A strong or weak foliation is accompanied in such rocks by a high or low mica content, respectively. Following this rule, a gradual transition between mica-rich and quartz-rich layers can also be handily represented by using curved lines.

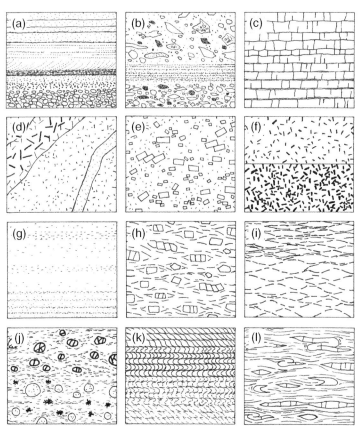

Figure 3.4 *Schematic sketches of various rocks. Original size of the entire sketch A4; felt-tipped pen with a usual line weight of 0.1 mm, occasionally ranging to 0.5mm in certain cases; natural size of the rock areas: (a) to (d) ca. 1 m, (e) to (m) ca. 10 cm. **(a)** Sandstone layers of different grain size; from bottom to top: conglomerate, coarse, and fine gravel (only indicated by light dotting); a thin clay layer (represented by short strokes); cross-bedded sandstone (lamellae indicated by dotted lines); top layers: coarse sandstone (indicated by light dotting); solid lines: thin clay layers separating and, consequently, defining the sandstone bedding. **(b)** Two tephra layers of different composition, size and shape of fragments, and grain size of groundmass, separated by a tuff layer with internal layering structure (dotted lines); tephra groundmass represented by dotting.*

Because wide-spaced lines depict a higher quartz content, the drawing often appears lighter in those areas. This matches the brightness of quartz-rich rocks in nature, in contrast to the quartz-poor schist, which is often darker.

These examples are intended only as anecdotal evidence as to the way in which rocks can be represented in meaningful drawings. Since rocks—even if they have the same name—can have, in part, significantly different structures, their representations may also differ significantly. It is only important that certain rules are followed to guarantee the recognition of the rock and its structures in the drawing. Strong contours of structures must be translated into correspondingly

---

Figure 3.4 *(continued)* *(c) Bedded limestone with cross fractures and thin clay interlayers. The fractures are limited to single layers and characterize the limestone. (d) Homogeneous granite with a thick pegmatite and a thin aplite vein. The short strokes represent (i) biotite platelets in granite, (ii) biotite and/or crystal faces in aplite, and (iii) crystal faces and feldspar cleavage planes, respectively, in pegmatite. These various types of strokes sufficiently and precisely represent the homogeneity and grain fabric of the rocks and the random orientation of biotite. (e) Porphyritic granite with flow fabric. The feldspar phenocrysts are schematized as rectangles and the biotite platelets as short dashes. These dashes also accentuate the isotropy of the groundmass. (f) Diorite (top) and gabbro (bottom) characterized exclusively by short, thin and short, thick strokes representing amphibole and pyroxene crystals, respectively. All other minerals are left out for clarity. (g) Mica quartzite. Only mica platelets are shown as short strokes. They indicate bedding with various mica content and the schistosity (grain orientation). (h) Augengneiss with schistosity represented by the alignment of biotite platelets (short dashes). Deformation under greenschist-facies conditions or lower temperatures is indicated by brittle behavior of feldspars (cross fractures in phenocrysts). (i) Augengneiss with conjugate foliation planes which are delineated by aligned biotite (short dashes). The biotite platelets frame lenses of plastically deformed feldspars, thus pointing to deformation temperatures higher than those of greenschist facies. (k) Garnet mica schist (top) and chlorite albite schist (bottom). Garnet and albite blasts are optically separated by different line weights and internal fabrics (fractures in garnet and small mineral inclusions in albite). Chlorite occurs in rosettes. The sigmoidal conjugate foliation planes are formed by mica platelets (short dashes). (l) Crenulation cleavage of different intensity, represented by solid lines (top) and broken lines (bottom). (m) Mica schist with quartz veins that are oriented parallel to the first schistosity and were isoclinally folded during the second deformation event. Typically, the quartz veins are dismembered into lenses and isolated fold crests. The cross fractures are characteristic of such quartz veins and facilitate their identification in the drawing. In addition, only few foliation planes are sketched as longer strokes. For clarity, the groundmass is not illustrated.*

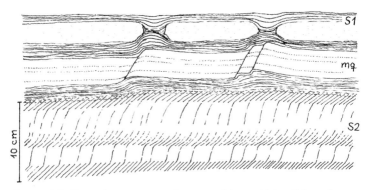

Figure 3.5 *Alternating sequence of metamorphic cm-thick pelitic and psammitic layers affected by two deformation events, which led to two schistosities, layering-parallel S1 and transverse S2. Lower part of the drawing: cm-thick layers with variable mica and quartz content, which, accordingly, show different intensities of foliation. The gradation of mica content leads to variation in orientation of the foliation planes. Upper part of the drawing: a mica-quartzite layer (qm) with weak layering between schist layers is fractured. In contrast, smooth boudins are developed in a fine-grained quartz-mica layer represented by fine dotting. These structures indicate the relatively higher strength of the mica quartzite.*

thick lines in the drawing. Only weakly recognizable boundaries, foliations, etc. are represented with thin or broken lines. Mineral typical forms and internal fabrics should be taking into consideration in the drawing. The orientation of crystals in rocks must match that of crystals in drawings. A light rock must generally appear light in the drawing and vice versa. Light and dark can be exaggerated in either direction to increase the contrast, and thus the readability, of the drawing.

Almost all of the drawings in this chapter are not labeled. This need not be so. The label-free drawings only demonstrate how well one can pass up labels in many cases and how meaningful just the line drawings of rock structures can be. Of course labels should be used if they serve to clarify the representation or if structures are ambiguous without them. Succinct abbreviations are, for the most part, sufficient. Many drawings with labels that are necessary, because they serve a faster and better understanding, will be shown in the next chapter.

3.4 Development of Drawings

Regardless of whether one is sketching rock structures from a cut sample on the table, drawing from an outcrop in the field, or compiling a drawing from multiple observations in a quarry, this drawing always undergoes a development. Logically, this development follows rules—the most important being: from large to small.

First, larger structures are drawn (rock layers, dykes, folds), followed by striking internal structures (Figure 3.6a). If at this stage another structure is discovered, it can easily be added by extending the drawing in either direction (Figure 3.6b). Such extensions occur mainly while drawing in the field when one walks along an outcrop and discovers something new. It contradicts all experience that after a while of looking and surveying, all important rocks and structures can be fully recognized and drawn in one continuous process. Since drawing is dependent on observation and a part of it, the interplay between drawing and observation leads to a stepwise development of the drawing. Therefore, it is advisable not to start and squeeze drawings in one corner of the drawing paper but, instead, to always leave enough space for additions.

In the course of this development, more structures will be added in the next steps as they are discovered—like further generations of foliations in a metamorphic rock, for example (Figure 3.6b). The next steps of drawing are about (i) adding internal fabric and filling layers (e.g., dykes) to better characterize the rocks and (ii) labeling and—if necessary—including a scale (Figure 3.6c). This scale, however, will rarely apply to all parts of a drawing. Typically, it applies to the bigger structures, like the thickness of layers or the wavelength of folds. The rock fabrics, and even more the internal fabrics of mineral grains, are often, necessarily, depicted in significantly enlarged form. In reality, the distances between foliation planes, for example, is about 1 to 2 mm; in drawings, this distance is about 20 times greater. Likewise, the mineral blasts that show up in some layers are also enlarged. In case important internal fabrics are too small to depict in the drawing, if can be useful to magnify and depict them separately. This often concerns the internal fabric of mineral grains. But details of deformation fabrics can also be important in their enlarged version for the geological message of the drawing.

The point of the drawing is not to correctly map the number of foliation planes or mineral blasts per meter or per area. This would not be a meaningful geological message. A useful piece of information, however, is the density of foliation planes, as this says something about the deformation intensity and about the mica-content of the rock. Even the number and size of folded quartz veins are rarely relevant, but their lenticular deformation and tight folding are certainly of importance. Since such quartz veins typically form parallel to the first foliation, isoclinal folds and the lenticular deformation point to an intense deformation during the second phase of deformation. In all five "complete" isoclinal quartz vein folds, long, short, and long limb follow clockwise after another. This can be taken as evidence of a clockwise shearing during the second deformation phase. Even from the position of the third foliation relative to the first and second, a clockwise shearing can be derived.

In garnet-mica-schist, short lines on the blasts indicate that the garnet was enveloped by the schist, so, it is older. Since the relationship between feldspar

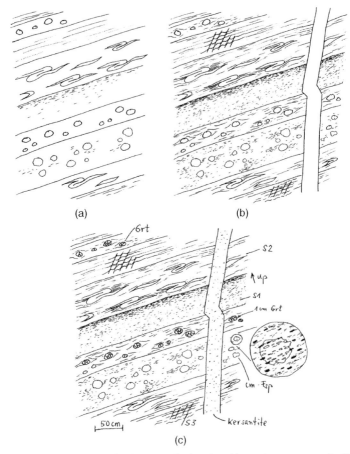

(a)

(b)

(c)

Figure 3.6 *Schematic development of a drawing. Alternating sequence of schist of variable mineralogical composition and a transverse young vein. (a) Sketch of the layering and its slightly inclined orientation. The horizontal in the field is equivalent to the "horizontal" in the drawing. Distinctive internal structures are integrated: the main schistosity—as long or short strokes depending on the schistosity intensity; isoclinally folded quartz veins that, in schist, are usually well visible; mineral blasts (garnet and feldspar) with slightly different line weight (higher for garnet, lower for feldspar).*

Figure 3.6 (continued) (b) A vein detected during this stage of drawing is added to the rightward extended drawing. The schistosity, also detected late, and the weak folding between foliation planes are superposed on the main schistosity in certain parts. The main schistosity is locally complemented, in order to visualize grading in one of the layers, and is intensified along the isoclinal folds of quartz veins to increase the optical contrast. (c) Garnet blasts are differentiated from the feldspars by their internal fracture pattern. The fine grain fabric of the vein is represented by light dotting. Labeling with typical abbreviations improves the readability of the drawing. Last but not least, the enlarged details of the internal fabrics of feldspars serve for better presentation of the deformation processes. The scale only roughly indicates the thickness of layers and the vein. It is obvious that the scale is not valid for the other structures, such as the sizes of mineral blasts or the spacing of foliation planes.

blasts and schistosity cannot adequately be represented, even with enlargement of the internal fabric, the information that the feldspar blasts overgrow the foliation and are therefore younger is delivered in a separate enlargement of a small rock section. A separate scale for this enlarged detail is not necessary. The size of the blasts is not an essential piece of information—rather the particle size distribution, which cannot be adequately represented in such a drawing anyway.

The few internal structures are sufficient to (i) characterize rocks, (ii) reflect the different degrees of deformation, (iii) clarify the shear sense during the second and third deformation phase, (iv) depict relationships between deformation and mineral growth, and (v) provide specific additional information like the direction of the stratigraphic top, for example. The labeling supports this characterization, but is not necessary in all cases. The structures shown in the drawings are easy and basic. Being able to recognize them in the field, on the rock, requires no special knowledge or exceptional observational skills. Many fabrics, like the folded and lenticular quartz veins, are so striking that they can be drawn without understanding their meaning.

3.5 Illustration on Different Scales

Drawings of rocks and their structures can be customized according to many scales—in extreme cases, from the micrometer all the way to the kilometer range. Drawings in the micro-range are easier to make, since the structures are usually formed more clearly and have no gaps. In thin section, structures are completely exposed. On a large scale, boundaries and contrasts frequently become blurred. In overview drawings, of quarry walls, for example, rocks can often no longer be characterized by their internal structures. Here, light-dark provides the only possibility for distinction.

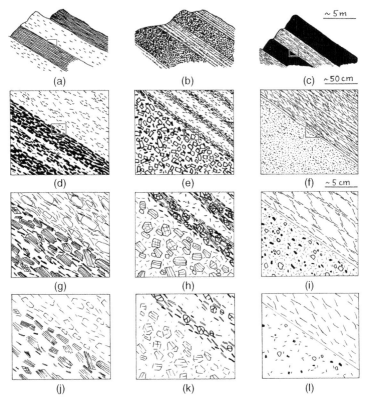

Figure 3.7 *Schematic representation of rocks on three different scales. The boxes mark the previous sketch. Size of the original drawing ca. 16×21 cm; felt-tipped pen with variable line weight. (a) Meter thick alternating layers of strictly banded amphibolite (dark) and clearly deformed augengneiss (light). (b) Meter thick layers of gabbro (dark, dotted) and granulite (striped). (c) Meter thick layers of basalt (black) and schist (striped). (d) Detail of (a) with centimeter to decimeter thick layers in amphibolite and gneiss. In comparison to these layers, amphibole (thick, short strokes) and biotite representations (thin, short strokes) are not to scale. (e) Detail of (b) with schematized and not-to-scale pyroxene (open rectangles) and biotite (short, thick strokes) in gabbro, and with mafic portions in granulite (strokes of different length and thickness).*

Once a single layer in a drawing is a few millimeters thick, rocks can be characterized on the basis of their internal structures. This is easiest for rocks that have typical structures. Good examples of this are the mafic plutonic, volcanic, and metamorphic rocks that have similar chemical and mineralogical compositions but distinctly different structures. Layered amphibolite is best depicted with a narrow stripe pattern (Figure 3.7a), gabbros through a coarse pattern of dots, which emphasizes the randomness of the structure (Figure 3.7b), and basalt through complete blackening (Figure 3.7c). Lighter rocks require an optically lighter internal fabric. On a large scale, the map symbols can be similar, like the "coiling" of orthogneiss, for example (Figure 3.7a). Foliated rocks are best characterized through parallel lines of various types (Figures 3.7b and c). Of course, stripes or "coil-symbols" don't represent the layering of amphibolite and the "softly" deformed feldspar augen of gneiss to scale. In a next enlargement step, the internal fabrics of the different rocks can be shown in more detail (Figures 3.7d-f). Now, it is possible to depict not just the layering but individual mineral grains in their characteristic outlines. Even the representation of preferred orientations and typical configurations of foliation planes (e.g., S-C fabric) is possible. At this magnification, the light-dark contrasts, as opposed to the larger-scale representation, are reversed in part. In the next and last step of magnification, a scale is reached in which even the internal fabric of the minerals and the details of the rock fabrics can be represented (Figures 3.7g-i). Larger structures, however, are lost at this scale.

The "optimal" filling of such drawings depends not only on the time available, but is primarily determined by how many details are necessary for the meaningful "geological message." We are usually inclined to fill our drawings too much. Omission is better and "less is more." It is up to the viewer to decide whether the structures in Figures 3.7k-m appear too sparse or not.

---

*Figure 3.7 (continued) (f) Detail of (c). Dot- and stroke-based schematization of basalt fabric and half-schematic representation of S-C fabric in schist. On this scale, the dark-light contrast between these rocks is inverted compared to (c). (g) Detail of (d). Amphibole in amphibolite and feldspar in gneiss are schematized based on line weight, grain shape, and internal fabric, following **Figures 2.1** and **2.2**. (h) Detail of (e). Pyroxene in gabbro, and garnet and sillimanite in granulite are schematized based on **Figures 2.1** and **2.2**. The isotropic "groundmass" in gabbro and in the light layers of granulite are represented by short strokes. (i) Detail of (f). Phenocrysts in basalt (e.g., olivine, pyroxene) are sketched as open or closed polygons. The isotropic groundmass is represented by short strokes. Short strokes divide the distinct foliation planes of the S-C fabric in schist into their biotite and white mica components. (k, l, m) Schematized rock representations of (g), (h), and (i), with more intense omissions, that is, with generally fewer strokes.*

The geological objects that we sketch in nature usually range across several orders of magnitude from the centimeter- to the outcrop-scale. This influences the schematization of the structures. A gneiss structure looks different in the centimeter-scale than in the meter-scale, a specimen of a conglomerate layer different than in the outcrop face. Nevertheless, the rocks and their characteristic structure should be easily recognizable at any scale—and, if possible, without complicated and extensive labeling. This problem can be tackled in four different ways and their combinations. (i) Characteristic signatures, with approximately three to four increments, are conceptualized for each scale. Thus, for every rock and for each of the main structures (foliations, layering, etc.) there are three to four typical forms of representation. (ii) Typical structures are largely drawn independent of the scale. Thus, in cm-, dm-, or multiple meter-thick layers, the phenocrysts of porphyritic granite will always have the same size and shape. (iii) If the drawing must span multiple orders of magnitude, for example, encompassing an entire outcrop face, the larger structures are drawn more or less to scale and supplemented with magnified details of smaller structures. (iv) Uncharacteristic structures can be explained with the help of labels. In practice, a compromise between these four options has been tried and tested. Typical structures are in part adapted to the scale, in part drawn independent of the scale, and supplemented with detail magnifications and (sparing) labels.

3.6   Detailed Drawings of Sample Cuts

Photos of samples or scans of sample sections are often dark or low-contrast. It is difficult to recognize the details of the rock fabric in them. The problem can be resolved with a drawing. This has the advantage that it is not only possible to enhance contrasts, but that the important structures can be emphasized by omitting the unimportant ones. In a publication, such a drawing is an excellent alternative to a photo.

Even in this type of "copying," the usual rules apply: The drawing consists only of black lines and dots. Various concentrations of lines and dottings are used for contrast enhancement. Dotted lines appear "lighter" than solid ones. The drawing is not filled but, rather, visually relieved by gaps. We leave it to the brain to supplement these gaps. After all, it is ideally suited for such a task!

In contrast to a quick sketch in the field, in this type of drawing, we utilize the time and peace at hand as well as the technical possibilities to copy and draw the rock structures precisely and "one-to-one." This means, the sizes and ratios are preserved. Even now, drawing pays service to observation but much more still to the exposing of details. This is not possible when in the field.

The more contrast the structures have, the easier they are to draw. The folded metapelite depicted in Figure 3.8 contains mm-cm-thick quartz veins whose white contrasts well with the rest of the dark rock. Following the outlines of the scans,

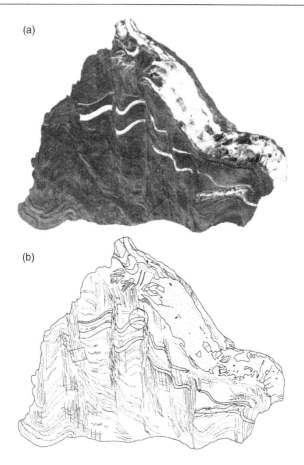

(a)

(b)

Figure 3.8 *Low-metamorphic metapelite, foliated and folded during the Variscan orogeny, with mm- to cm-wide quartz veins (Middle Rhine Region near Loreley, Rhenish Massif, Germany).* **(a)** *Scan of a sample section perpendicular to the fold axis; long side of sample ~ 15 cm.* **(b)** *Drawing of the section with only locally indicated structures; drawing on tracing paper placed on a scan print; felt-tipped pen with line weights of 0.1 to 0.2 mm; quartz veins are represented as slightly dotted, bedding and first schistosity as dotted lines, and the second schistosity as continuous lines. Locally, the quartz veins diverge from the bedding planes (circle). The second schistosity, transverse to bedding, is intensified in shear zones (rectangle).*

it is the white layers that are best sketched first. This gives the drawing a rough structure. The small shear zones that intersect the layering follow next and then the individual planes of the foliation. Lastly, the layering that is barely visible in the scan is implied with dotted lines—preferably only so many that the light folding is visible without filling the drawing too much. The dotting is meant to visually highlight the quartz veins. An alternative would be to intensify the foliation and layering around the outside of the quartz veins. But this could likely fill the drawing too much. A label could help to explain various structures but is really not necessary—with the exception of the veins. These can be made of quartz or calcite or both, which cannot be discerned by simply looking at the drawing.

Sketching rocks with diffuse structures is more difficult and requires more effort (Figure 3.9). If time is of the essence, the drawing is reduced to the distinctive boundaries. These can appear as solid lines or as short lines in diffuse areas (Figure 3.9b). One stops drawing as soon as the essential structure (in this case, the fold) is visible. The fold appears even more clearly in a pure line drawing (Figure 3.9c). Here, boundaries are "interpretively" drawn in diffuse areas even where they are not visible in the rock. Dark components are represented especially at the borders to the light layers. On the one hand, this increases the contrast and, thus, the readability of the drawing; on the other hand, the short strokes, with which the biotite platelets are marked in this case, provide the possibility of clarifying the details of the foliation, and with that, enhancing the geological message of the drawing. The drawing that most closely models nature is one in which the boundaries are forgone and the layers are exclusively highlighted

---

Figure 3.9 *Folded biotite-feldspar gneiss from the basement of the eastern Tauern Window (Eastern Alps, Austria); section perpendicular to the fold axis; short side of section ~ 10 cm; light = quartz-feldspar layers with mm-grain size; dark = biotite-rich quartz-feldspar layers; different graphical representations of the scanned sample section; drawings on tracing paper placed on a scan print; felt-tipped pen with line weights of 0.1 to 0.5 mm. (**a**) Scan of the sanded, but not polished, section. The light feldspar layers are partly sharply confined by dark biotite layers. The changes from feldspar-rich to biotite-rich layers are partly gradual. (**b**) Drawing of the boundaries between light and dark areas. (**c**) Drawing with more strongly schematized boundaries. Size and orientation of biotite platelets are indicated by short strokes. Thus, the folded first foliation and the locally developed second foliation are illustrated (rectangle and circle, respectively). (**d**) Drawing without boundary lines. The layers are represented and separated from each other solely by their different content of biotite. Thereby, even diffuse transitions can also be made visible.*

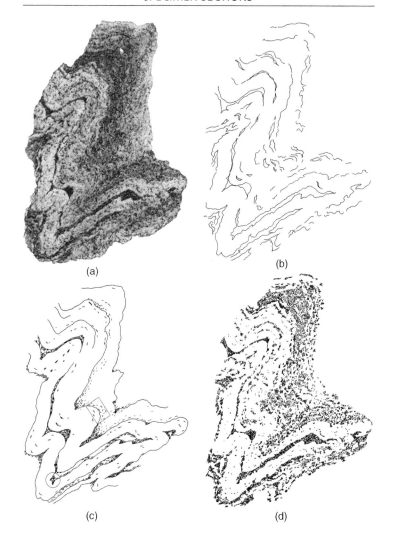

(a)

(b)

(c)

(d)

through the distribution of short lines (Figure 3.9d). Here, the natural structure can be depicted quite accurately through line length, thickness, orientation, and concentration. If the gradual transitions between certain layers are regarded as important, this type of drawing is surely the first choice. If, however, the type of folding and the schistosities are more important, a line drawing with dotted highlighting of the schistosity planes is sufficient (Figure 3.9c). This can be created with much less effort. As always, it comes down to: which geological message should be conveyed and how can this be done most effectively and with the least effort?

In the absence of larger, clear structures, the only remaining options are: "simple and fast, but distorted" or "time-consuming, but realistic." The sample section of a folded augengneiss depicted in Figure 3.10a can be converted to a drawing in which the feldspars are not represented by contour lines but by omission (Figure 3.10b). The groundmass of the rock is what is almost exclusively represented. With this, the drawing comes quite close to reality. The price, however, is the high time and work effort needed to create the drawing. To clarify the size distribution of feldspars and the diffuse foliation fabric with its small, local vortices, it suffices to draw the outlines of the feldspars and some of the larger biotite platelets. This saves a lot of time, though it is not so close to nature, but still conveys the message that matters most for this rock (Figure 3.10c).

Figure 3.10 *Weakly foliated and folded augengneiss from the Sesia Zone (Val d'Ossola, Western Alps, Italy). (a) Scan of a polished sample section; mm-cm-sized feldspars stick in a μm-mm-grained groundmass of quartz, feldspar, and biotite; short side of sample ~ 10 cm. (b) Drawing of the scan only representing the approximate size, orientation, and accumulation of biotite by short strokes; drawing on tracing paper placed on a scan print; felt-tipped pen with 0.1 to 0.2 mm line weight. (c) Drawing only representing the outlines of larger feldspars with few local biotite platelets.*

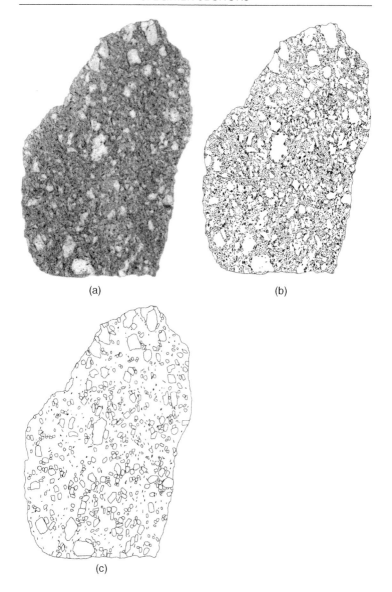

(a)

(b)

(c)

Summary

- Ideally, first observation and scientific interpretation, followed by the decision of what to draw and what not, should proceed drawing a rock section.
- In reality, observation is the true beginning. But it is also part of drawing, interacts with it, and persists throughout the duration of the drawing process.
- Not the true-to-life reproduction, but rather the transfer of scientifically relevant information is important.
- Strong contours of structures require thick lines in a drawing—weakly visible boundaries or foliations, thin or broken lines.
- The filling of structures must be adapted to the shape of the structure or neutral.
- The way in which rocks and their structures are schematized depends on the scale. Every scale requires a different kind of schematization.
- Typical rock schematizations with high recognition value, like the cross fractures in limestone or long, non-oriented lines for pegmatite, for example, can be used at all scales.
- Scale loyalty is usually neither desirable nor feasible, but the shapes of the internal fabrics should be correctly represented.
- Internal fabrics of rocks are predominantly distributed in a "fractal" fashion and should be drawn that way.
- When drawing the grain fabric, it is usually sufficient to draw only the dark minerals.
- For the representation of differently sized fabrics, it is advantageous to magnify small areas of the drawing and place them separately next to the drawing.
- Omission is (not only to save time) a good strategy while drawing. Drawings should stay "visually light." If one has the feeling of having drawn too little, it's probably just right.
- When time is limited, it is better to draw the structures in small areas accurately rather than to fill large areas inaccurately.
- Develop drawings gradually! First, large structures are drawn, then their internal fabrics and fillings; lastly, the drawing is labeled.
- Drawings grow. Fabrics that are discovered in the course of observation and drawing must be integrated or "added-on." Therefore, enough space should be left around the original drawing.

4

DRAWING ROCK STRUCTURES IN THREE DIMENSIONS

Sometimes rock structures are only recognizable on (often randomly oriented) surfaces and can only be represented flatly. Although important information is still retained and conveyed this way, as shown in numerous publications, the lack of the third dimension is still a shortcoming. Most rock structures are layered but always develop in three dimensions. Even linears have a spatial position and can only usefully be shown in three dimensions. This is why drawings in three dimensions, perspective drawings, are always preferred over two-dimensional representations. Outcrops in the field rarely exist as smooth faces. There are almost always cuts that are perpendicular, or nearly perpendicular, to one another; or one can take specimen sections and look at the structures in 3D. Even at just a few centimeters, there are surfaces on which 3D structures can be identified. Even on smooth quarry walls, a hammer and chisel can be used to expose small steps in which 3D structures become visible.

Beautiful and instructive 3D sample drawings can be found in numerous, specifically older, publications. Many of them are scaled but a few are unscaled and schematized. As a further development, the generally unscaled drawings of geological structures (samples and outcrops) that yield a much higher information content through the shifting or lifting of individual parts and continuing of structures towards the outside, were introduced by Gerhard Voll in the early 60s (Voll, 1960, v. Gehlen & Voll, 1961, Nabholz & Voll, 1963) and taken up by other authors, albeit sporadically, in later years (Steck, 1968, Steck & Tièche, 1976, v. Gosen, 1982, Kruhl, 1984a, Vogler, 1987). They provide a significantly deeper insight into the structure and structural development of rocks and regions.

While drawing structures in three dimensions, there is plenty of room for different representations and freedom to develop a signature style. Nevertheless, 3D drawing follows the same rules that apply to the drawing of 2D structures. Furthermore, it is useful to take some additional rules into consideration that make the drawing process easier, ensure good interpretability, and also serve the aesthetics of the drawing.

In the following sections, we will first touch on some basic rules, then look at how to best develop 3D drawings, compare field book sketches with "clean drawings," and, finally, discuss some exemplary drawings.

Drawing Geological Structures, First Edition. Jörn H. Kruhl.
© 2017 John Wiley & Sons Ltd. Published 2017 by John Wiley & Sons Ltd.

4.1   Foundations

For field drawings, a drawing area of about 15 x 20 cm (A5), that can be enlarged to about 20 x 30 cm (A4), should be available. This does not mean that the drawing itself must be this large. If only a few rock structures are present or need to be drawn, a smaller drawing is sufficient. For complex, 3D drawings with detailed labeling, at least an A5-sized area is usually necessary. When the area is smaller, structures and labels are too close together and the drawing's readability is compromised. This also means: a field book should not be much smaller than A5. This size is also convenient because it still fits in a bigger jacket/pants pocket or belt bag.

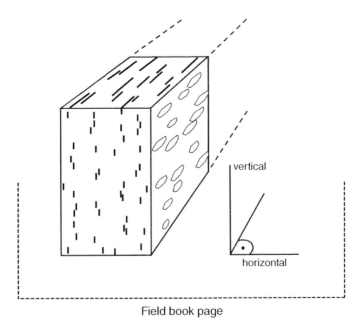

Field book page

Figure 4.1 *Rectangular block of gneiss with one surface frontally facing the viewer and two side surfaces perspectively extending backwards (broken lines). Mica platelets are schematized as shorter and longer strokes. The mica orientation defines a schistosity perpendicular to the front and top face, and parallel to the lateral face. The lineation, developed on the foliation plane, is marked by the long axes of elongate and aligned micas, and runs parallel to one edge of the block. The horizontal and vertical edges of the front face are parallel to the margins of the field book or canvas. Thus, these edges are defined, or perceived respectively, as horizontal and vertical in nature.*

118

As stated in Chapter 1, outcrops or rock samples are not drawn strictly as found in nature but schematically, to highlight the important details and omit the unimportant ones. This means that schematizing the outline of the sample is the first step. The easiest form of schematization is a rectangular block in which the planar rock structures are parallel and perpendicular to the block surfaces and the linear structures are parallel to the block edges and perpendicular to the block's surfaces (Figure 4.1). The front block surface either directly or nearly faces the observer. Two other surfaces extend "diagonally backward." This ensures that the structures are not only visible from the front surface but also recognizable on two further surfaces as well as recognizable in their 3D orientation.

Starting from the front surface facing the observer, the block can either be drawn to the back right or the back left (Figure 4.2a). This distorted perspective benefits the spatial appearance of the sketch. The distortion need not be strong. It is enough if the backward-extending edges of the block eventually meet at a vanishing point far in the distance. If there is no surface, but rather an edge of the block, facing the viewer and the block faces point away from the viewer, there are two distant vanishing points (Figure 4.2b). This block orientation has the advantage that the three visible surfaces are approximately the same size and structures can be represented on them in a clearly visible manner.

Since the drawing paper or field book is usually held so that the two edges are oriented "horizontally" and "vertically," one instinctively adopts the same horizontal and vertical edges of the drawing in the field. This means: edges of the 3D

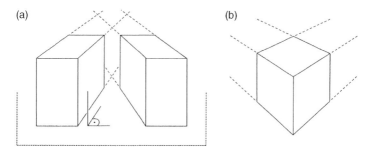

Field book page

Figure 4.2 *Rectangular blocks as basic elements of a 3D drawing. (**a**) Blocks with a weak perspective distortion toward the "right back" and "left back." The prolongations of the edges (broken lines) meet at remote vanishing points. The edges follow the sketched rectangular coordinate system. (**b**) Block with one vertical edge facing the viewer. The prolongations (broken lines) of the horizontal, backward-running edges meet at two different, remote vanishing points. All three block surfaces are equally well visible.*

block that are oriented parallel to the edges of the drawing paper are interpreted as horizontal and vertical. If a steeply plunging foliation is represented in a block of this orientation, it cuts through the side faces of the block (Figure 4.3a). This leads to arbitrary cut lines. Moreover, the surface of the foliation is not visible and the structures on it (a lineation, for example) cannot be represented properly. To avoid this, the block is inclined until one side surface is parallel to the foliation (Figure 4.3b). The lower edge of the field book remains the marking line for the horizontal. The block edges extending "diagonally backward" are perceived as being horizontal.

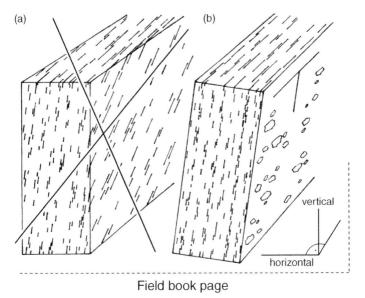

Field book page

Figure 4.3 *Rectangular blocks (slightly perspectively distorted) with foliation planes that are represented by aligned mica platelets (short strokes). **(a)** The steeply inclined foliation planes run obliquely to the block and generate random, geologically insignificant cut lineations on the tabular side face. This is an unacceptable drawing! **(b)** The rectangular block is inclined to such an extent that the schistosity is oriented parallel to the lateral faces and perpendicular to the front face. The lineation (elongate mica platelets), related to the schistosity, is shown on the lateral faces of the block. The horizontal line of the T-sign symbolizes the horizontal on the foliation plane, and the longitudinal one the dip line. Two axes of the rectangular coordinate system follow the edges of the field book. The third, backward-running axis completes the system in 3D. Accordingly, the backward-running edges of the block are regarded as horizontal.*

Generally, as far as it's feasible, the faces of the block should be parallel and perpendicular to planar and linear rock structures (layering, bedding, foliation, lineation, etc.). Regardless of the edges of the block, the horizontal and vertical planes in a drawing can be specified with a T-sign (Figure 4.3b), which also serves, on geological maps, as a symbol for the strike and dip of a layer.

When dealing with folded layers, it is usually impractical to use the block as a sketch's base form. Even if the side faces of the block are oriented perpendicular and parallel to the fold axis, they do not provide views of the folded layers (Figure 4.4a). It is better to move away from the rectangular block shape and

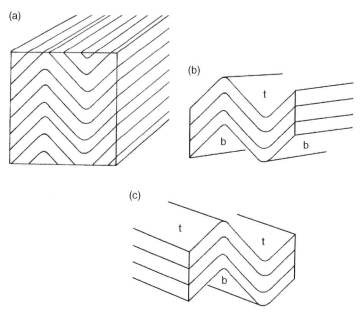

Figure 4.4 *(a)* *Rectangular block cut out of folded layers. The orientation of the fold axis can only be derived indirectly from the intersection lines of layers with the lateral block face. Structures on the layers cannot be illustrated. Such a sketch is nearly useless!* *(b)* *Sketch of the same folded layers but 'broken off' the block. The orientation of the front face is unchanged, whereas the right lateral face is rotated clockwise around a vertical axis. The bottom side of the lowermost layer (b) is similarly well exposed as the top side of the uppermost layer (t).* *(c)* *With the same orientation of the front face, the left lateral face of the folded layers is rotated counterclockwise and, thus, becomes visible. The top side of the uppermost layer (t) is well visible; the bottom side of the lowermost layer (b) less so.*

adjust the 3D drawing to the fold form. In the first step, similar to drawing a block sketch, one draws a surface that corresponds to a section that is perpendicular to the (horizontal) axis of the folded layer and that faces the viewer (Figure 4.4b). The layer is cut perpendicular to its planar extension. The cut edges extend towards the back left—that is, the "horizontal" direction in 3D. These cut edges serve as reference lines for the fold axis. If the axis is horizontal, it is drawn parallel, as in this example, to the cut edges. In such an orientation, the folded layers of the upper and lower sides are almost equally well visible. But, if there are important structures on the upper side of the top layer that should be easily recognizable, the fold must graphically be turned counterclockwise until these surfaces are sufficiently visible (Figure 4.4c). Despite the rotation, the "horizontal-vertical" reference system does not change. The reference system also allows for the dip of structures (e.g., fold axes) to be visible (Figure 4.5). The dip angle of the fold axis can be estimated by the angle between fold axis and the "backward" extending edge of the upper cut-off layer.

Drawings of echelon blocks, in which layers are differently truncated or shifted against each other (Figure 4.6), allow for an even better insight into structures. Each layer now offers views from three orthogonal surfaces. 3D structures can be optimally represented this way. This applies to structures that are only visible on a layer, like lineations, and also to structures that only become apparent in three dimensions, like different kinds of cross bedding. If the front surface of the

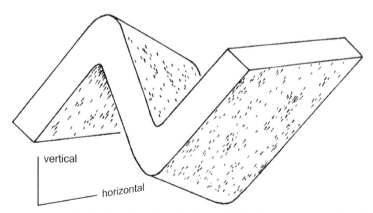

Figure 4.5 *Sketch of a fold with inclined axes. The backward-running edges of the folded layer mark the horizontal. Accordingly, the fold axes appear at a ca. 40 to 50° incline to the horizontal.*

(a)   (b)

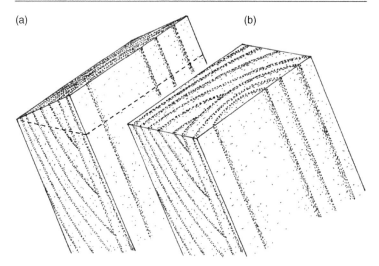

Figure 4.6 *Layer in different orientations. **(a)** Inclined layer with internal cross bedding. The top face is barely visible. **(b)** The top face is slanted relatively to the layer and, thus, shows a clearly increased area with visible structures.*

echelon block is to remain facing the viewer, the "upper" of the two cross sections of flatly dipping layers is often so thin that structures on it are not easily visible (Figure 4.6a). Cutting the layers on a slant, thus, increasing their cross sectional area and making them parallel to the horizontal planes, is a simple remedy (Figure 4.6b). This also means that an echelon block with numerous plates generally extends transversely to the strike of layered structures. This is how the echelon block represents a three dimensional block-cross section through these structures.

Three orthogonal cuts perpendicular and parallel to the most important planar or linear structures provide maximum insight into the rock structure. It follows that differently oriented geological structures should not be depicted in compact blocks if possible, but rather composed with plenty of space in between (Figure 4.7). Even if this makes completing the drawing somewhat more demanding, it is well worth the effort. It results in drawings with many surfaces that offer a lot of information about the rock structure. It doesn't matter whether the geological structures are visible in this exact form in an outcrop or sample. "Lifelike" drawing is not the goal and also does not make sense. The aim is to generate clear drawings that provide maximal information with minimal effort.

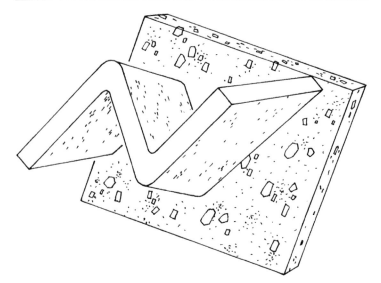

Figure 4.7 *Two different geological structures (fold and transverse porphyritic vein) assembled to an open structure that leads to maximum visibility of relevant surfaces (surfaces of vein and layer) and structures (orientation of feldspars [boxes] and micas [short strokes]).*

Principles of Geological 3D Drawing

- The drawing of a sample or an outcrop is created in 3D. In it, the essentials are highlighted and the inessentials are omitted.
- The usually random and irregular surfaces of a sample or outcrop have no meaning and are not drawn.
- The simplest form of 3D drawing is a rectangular block, whose front surface faces the viewer and whose side surfaces are rotated so far that the structures on them are clearly visible.
- The surfaces of the block are parallel or perpendicular to the planar or linear rock structures if possible.
- The orthogonal edges of the field book page are the reference lines for the horizontal and vertical axes in the field drawing. In addition, a T-mark can be used as a horizontal and dip line on a surface.

- Layers are graphically displaced from one another ("echelon block") to make structures more visible in 3D.
- Rocks or outcrops with complex structures are not represented in compact blocks but, rather, in drawings with many surfaces, which provide maximum insight into the rock fabric.

4.2   The Development of Schematic 3D Drawings

4.2.1   Simple Blocks

As discussed in Chapter 4.1, a schematic 3D drawing, in its simplest form, is represented as a rectangular block. If the rock fabric is anisotropic, the block should be orthogonal to the anisotropy planes (bedding, magmatic or metamorphic layering, foliation) or anisotropy directions (lineations, fold axes). In the first step, the block edges and layer boundaries, and perhaps also the cardinal points for the orientation of the block, are drawn (Figure 4.8a). If it is foreseeable that all three orthogonal surfaces will have structures worth representing, the block should be rotated so that all three surfaces are equally well visible. In this position, the three "front" edges meet to form three 120° angles. The front vertical block edge is parallel to an edge of the canvas or field book and is thus perceived as the geographically vertical axis. This helps to ensure that the viewer perceives the upper surface of such a sketch as nearly horizontal and the two side surfaces as nearly vertical.

If the layering or other planes are inclined against the vertical axis, the block must be inclined as well so that the block surfaces remain parallel to these planes. The tilt of the block is most easily achieved by tilting the visible side surface and, concomitantly, the two front vertical block edges as well (Figure 4.8b). Now, the block surfaces are filled, step-by-step, with structures (Figure 4.8c). It does not really matter in what order this is done. It is, however, useful to first look at the distinctive and somewhat darker layers (orthogneiss and amphibolite, in this case), because they determine the appearance of the drawing. This way, if something cannot be drawn because of lack of time, at least the characteristics of the structures remain visible. In addition, one maintains the freedom to vary the contrast to the remaining layers through stronger or weaker fills. Generally, the coarse structures should be drawn first as they will be partially filled again.

The tilt of the block can also be emphasized through a strike-dip sign on the corresponding side surface. When the strike and dip values are noted on this sign, the exact spatial orientation of the block can better be represented than is possible through the drawing alone. The strike-dip sign also offers the possibility

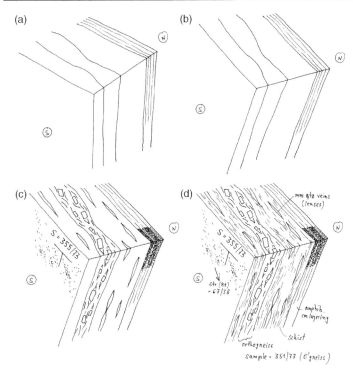

Figure 4.8 *Stepwise development of the schematic drawing of an alternating sequence of orthogneiss, schist, and amphibolite illustrated in four steps.* (**a**) *Block with outlines, layer boundaries, and cardinal directions for geographical orientation. The block is oriented in such a way that three orthogonal sections parallel and perpendicular to layering and schistosity, respectively, are well visible.* (**b**) *The inclination of the layers relative to the horizontal requires an equivalent inclination of the block.* (**c**) *First, additions are made: quartz lenses in schist; feldspars and biotite platelets in orthogneiss; the lineation in form of accumulated and oriented biotite platelets (thin short strokes); and amphibole crystals (thick, short strokes) depicting the variable amphibole content in distinct amphibolite layers. The T-sign, indicating the strike and dip of the layers, and the measurement of the schistosity are inserted relatively early, before the foliation plane is filled with strokes.* (**d**) *Further steps include: (i) representation of foliation planes in the schist by strokes of variable length, (ii) addition of the sample orientation, and (iii) labeling of the drawing. The relatively time-consuming filling of the amphibolite layers is skipped. Original size of drawing ca. A5; felt-tip pen with line weight 0.2 and 0.3 mm.*

of specifying the orientation of a lineation (Figure 4.8d). In the final phase, the foliation planes are filled (preferably "fractally") so that the type of foliation (relatively short planes, no metamorphic banding) is visible without losing the optical contrast to the gneiss and amphibolite layers. It suffices to fill the amphibolite layer in only a short segment and even the complete filling of the orthogneiss layer is not really necessary.

Labeling the individual structures and rocks complements and clarifies the drawing. It is up to each person to decide how extensive or sparing the labeling will be. For example, orthogneiss and schist (with the typical quartz vein lenses) are so striking in appearance, that labeling is not really necessary. On the other hand, important fine structures, such as biotite orientations, that mark lineations, for example, are hard to interpret without a label. The spatial position of the lineation should also be included in the drawing, even when the approximate slope of the lineation to the horizontal in the sketch can be estimated. The orientation of samples taken from an outcrop is also written directly next to the drawing.

In general, a drawing provides more information than is understood by the label alone. The strictly parallel planes in schist indicate a strong compression of the rock, the thin quartz lenses a deformation under conditions of quartz plasticity, that is, above about 300°C. The angular feldspar crystals in augengneiss, however, limit the deformation temperatures to a maximum of 500°C, the lower temperature limit for feldspar plasticity (Voll 1976a).

Slightly inclined or horizontal layers should be represented as horizontal plates (Figure 4.9). The schema by which the drawing is developed stepwise, always stays the same. A rectangular plate (Figure 4.9a) is cut in such a way that prominent linears and planes are parallel to the side surfaces of the plate (Figure 4.9b). If planes lie slanted to the sides of the plate, they are exemplarily drawn, that is, continued, from the inside to the outside of the plate. If the label is placed before filling the planes with internal fabrics, it can be tied more closely to the structures. The measured values of foliation planes, for example, can be written directly on the surface (Figure 4.9c). Only exemplarily drawing out of the plate makes it possible to represent structures that are developed on these planes (Figure 4.9d). This is also more elegant than cutting the rectangular block obliquely to obtain an additional surface. Some foliation planes and fractures are sometimes not discovered until the late stage of observation when the drawing is almost finished. Then, drawing the structure exemplarily is the only way of representing structures on these planes. It is convenient to bring young structures into the drawing as early as possible, because they often influence the orientation of older structures (the bending of older foliations around younger ones, displacement of layers at fractures, etc.). If younger structures are introduced too late, the drawing must usually be corrected later in unpleasant ways.

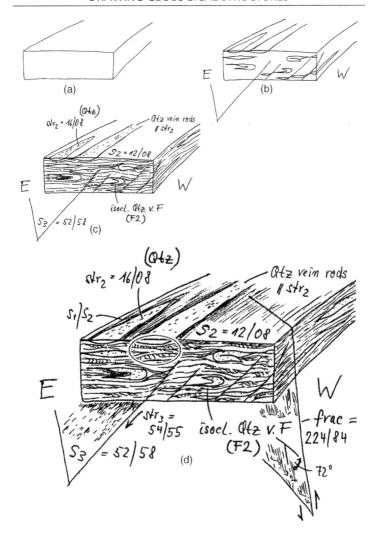

(a)

(b)

(c)

(d)

Figure 4.9 *If main structures are oriented horizontally, the largest block surface is oriented horizontally as well. (a) The outlines are represented slightly perspectively. (b) In addition to the integration of dominant structures and the main cardinal directions, the foliation planes that are not parallel to the lateral faces of the block are drawn exemplarily out of the block. Because older structures are often displaced by younger structures, it is advisable to draw these younger structures early on. (c) It is advantageous to add labeling and measurement values before filling with internal fabrics. This allows labels to be connected more closely to the structures they describe. The minerals that form specific structures (e.g., lineations) are indicated as necessary. (d) In the final step, (i) internal fabrics are drawn, (ii) missing labeling is added, and (iii) structures detected only during the late stages of observation are integrated or attached. In order to increase the contrast and, thus, the readability of the drawing, the internal filling close to pronounced structures (e.g., quartz vein rods in this drawing) may be intensified. However, filling too densely, as possibly in the present case, should be avoided. The drawing contains more information than is indicated by labeling, for example, relics of folded S1 between S2 foliation planes (ellipse). Size of original drawing ca. A6; ballpoint pen.*

4.2.2   Echelon Blocks ("Staffel Blocks")

A rock sample or outcrop is usually composed of different layers. If this is the case, it is useful to break up the drawing into individual, slab-like blocks that are offset from one another. This way, structures on layer surfaces, like lineations, for example, can be made more visible (Figure 4.10). This is essential, because linear fabrics in rocks are of great importance for the interpretation of geological processes. If there are no fabrics on the planes, they are only characterized by dotting or left blank altogether. Then, it suffices to make small areas of the planes visible (Figure 4.10a,b). The plates are, as usual, cut so that their orthogonal side surfaces are parallel or perpendicular to the planes and linears.

When striking fabrics, which need to be clearly shown, have developed, individual layers can be separated from one another in order to reveal larger parts of their surfaces (Figure 4.10c,d). It is important, however, that these "floating" layers retain their spatial orientation relative to the other layers.

Similarly, in the echelon blocks, the layers should also be inclined according to their spatial orientation. The direction from which the plates are viewed, should, in principle, not matter as long as the orientation is indicated by two given cardinal points. However, it is advantageous if the plates point towards rather than away from the viewer, so that all three orthogonal plate surfaces are clearly visible (Figure 4.11). In addition, the plates should all be about the same size. Even if this

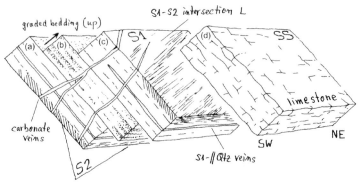

Figure 4.10 Schematic echelon block ("staffel block") representation of an alternating metapsammopelite sequence (a to c) of the Variscan basement with Mesozoic limestone cover (d). The single plates represent layers of different composition and equal orientation. The weak and bedding-parallel first schistosity, S1, is marked by few strokes (layers a and c) and a grading of variably dense dotting (layer b). In quartz-rich layers, a lineation is not developed on the foliation planes of the first deformation. Accordingly, these planes are evenly dotted. The second schistosity, S2, is only present in mica-rich layers (a and c), which, therefore, appear darker in accordance with their appearance in nature. A mica-rich region is intentionally placed in the upper part of layer (c), so that the intersection linear between first and second schistosity is visible on the surface. The layers are filled with internal fabrics only as far as necessary. The "floating" limestone layer leaves a large portion of the lower rock layer exposed. In addition, a several meter wide gap of exposure is indicated. The inclination of the echelon block towards the viewer provides equal visibility of all three orthogonal surfaces of the plates. Redrawn from field sketches; western termination of Monte Albo (Baronie, Sardinia, Italy); outcrop KR4652 (field book 39, Kruhl, 2002). Size of the original drawing A5; ballpoint pen.

Figure 4.11 Echelon block representation of a weakly inclined sequence of phyllitic and quartz-carbonate layers. Light-dark contrasts of the layers are generated by variably dense internal fabrics—basically foliation planes. The development of the drawing from top to bottom and from left to right can be deduced from the relationship between drawing and labeling. The upper layers were labeled before the right, backmost outline of the echelon block was drawn. After labeling, the second lowest layer was extended downward by a quartz-calcite vein and a phyllite layer. At present, its primarily plain boundary represents the upper surface of the quartz-calcite vein. The redrawing can be recognized by the slightly thickened line

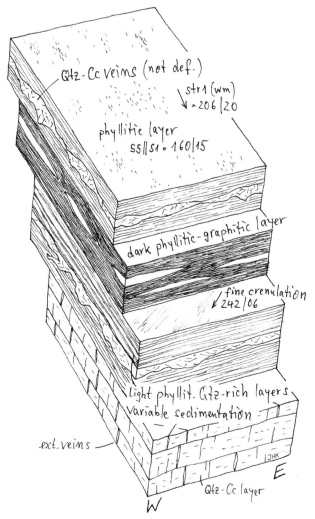

Qtz-Cc veins (not def.)

str 1 (wm)
↓ ≈ 206|20

phyllitic layer
SS || S1 ≈ 160|15

dark phyllitic-graphitic layer

fine crenulation
↓ 242|06

light phyllit. Qtz-rich layers
variable sedimentation

ext. veins

Qtz-Cc layer

W

E

Figure 4.11 *(continued) on the frontal face of the drawing (arrow). After labeling of the second lowest layer, the lowermost quartz-calcite layer was attached.
Redrawn field sketch (field book 35, Kruhl, 1994); labeling converted from German
to English. Outcrop KR4138; Valser Rhine Valley (Switzerland); size of original
drawing ca. A5; felt-tip pen with line weight 0.2 and 0.3 mm.*

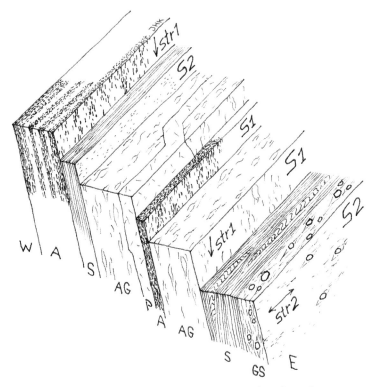

Figure 4.12 *Schematic echelon block representation of an alternating sequence of amphibolite (A), schist (S), augengneiss (AG), pegmatite (P), and garnet-schist (GS); felt-tip pen with line weight 0.3 mm and locally lighter due to obliquely set lead; size of the original drawing ca. A6. The layers are steeply inclined toward the east (E) and are oriented in such a way that the three orthogonal, lateral faces of the plates are equally well visible. The displacement of the layers increases the visibility of the parallel schistosities, S1 and S2, which bear the variably oriented lineations, str1 and str2. The second schistosity is sufficiently illustrated by thin layers with tightly folded planes of the first schistosity (arrow). Crystal-plastic deformation of feldspars is indicated by the lenticular shape of feldspars in the augengneiss (AG). This does not require additional labeling. To save time, some layers were left partially open.*

does not matter for the "geological message," it makes the drawing simpler and much clearer.

The echelon-block representation is of particular importance in a sequence of many different rocks with different structures (Figure 4.12). Regardless of its thickness, ever layer gets its own plate-like block. This way, all three orthogonal surfaces of the block are visible, especially the anisotropy planes and their fabrics. The plates need not be completely bounded and may even be kept open. Visually, this fits better to an incomplete filling of planes, makes the drawing "lighter," and saves time. However, the backward-retreating edges of the individual plates should be approximately the same length so that the 3D effect is preserved.

One usually (but not always!) has free choice when deciding on the order of plates. It is most clever to place the layer with the most variable fabrics, or the one that must be labeled most extensively and thus requires more space, at the start of the sequence (Figure 4.12).

When drawing in the field, it is usually best to start with an echelon block. It represents the basic form of the drawing that is expanded stepwise through progressive observation and the recording of rocks and structures, and can be combined with other forms, such as folds. This will be discussed in more detail later on.

4.2.3 Folds

The graphic representation of folds is a little more sophisticated than that of blocks but also follows simple rules. As described in Chapter 4.1, it is favorable to adjust the drawing to the fold formation. Since a sample or outcrop usually consists of several rock layers, the schema of the fold drawing is combined with the echelon block schema. The representation of an upright orthorhombic fold is easiest (Figure 4.13). First, a surface oriented almost perpendicular to the fold axis (which will later be drawn) is shown facing the viewer (Figure 4.13a). From there, the to-the-right- or backward-extending cut edges of the layers, marking the horizontal axes, are set (Figure 4.13b). With their help, the dip angle of the fold axis (or its horizontal orientation) is represented (Figure 4.13c). It is important that the cut edges of the layers extend backwards in such a way that the surfaces are clearly visible and internal fabrics (e.g., lineations) can easily be represented on them.

Sketching horizontal monoclinic folds works very similarly. First, a front surface, which represents a cut through layers displaced relative to each other and is (nearly) perpendicular to the fold axis, is drawn (Figure 4.14a). After this, the cut edges of the layers are extended obliquely backwards (Figure 4.14b). Then, the fold axes are drawn (Figure 4.14c). With the help of cardinal points (East = E and West = W, in this case), the orientation of the cut edges, and, thus, of the fold axes, can be defined.

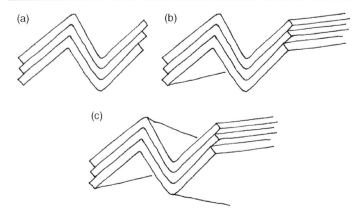

Figure 4.13 *Schematic drawing of an orthorhombic upright fold. The outlines adapt to the shape of the fold. Just as with the echelon block, the three layers are displaced against each other for better visibility of their flat faces. **(a)** First, the frontal face is drawn. It marks a geographically vertical plane and is cut approximately perpendicular to the fold axis. **(b)** The frontal face is supplemented by the edges of the layers, which run obliquely backward and mark the geographic horizontal. **(c)** The horizontal edges of the layers assist in representing the orientation of the fold axis, which has an approximately 20° dip in this case. When a drawing is oriented in such a way, the top face of the upper layer and the bottom side of the lower layer are well visible.*

It is more difficult to draw a fold when the axis is approximately traverse to the observer's viewing direction, as is the case when it's part of an echelon block that is oriented a certain way for other reasons, for example (Figure 4.15). A cut surface with a very acute angle with respect to the viewer appears to have a highly distorted geometry (Figure 4.15a). If the fold hinge faces the viewer, the curvature must be highlighted with lighting effects, that is, through differently dense dotting or line marking (Figure 4.15b). Both representations require care and an increased effort when filling areas. This is why one should avoid such fold orientations in drawings.

Drawing folded folds is a challenge, especially when the axes of the different fold generations cross one another (Figure 4.16). Older folds that bend around younger ones require a lot of attention making them very time-consuming to draw. Only with light-dark contrast can they be represented at least somewhat realistically in 3D (Figure 4.16a). Because this contrast is achieved by filling areas with internal fabrics (e.g., lineations) or with dotting, it requires a correspondingly large amount of time. Moreover, it requires a lot of caution to correctly represent the geometry of folded folds. As part of a time-saving alternative, folds with steep

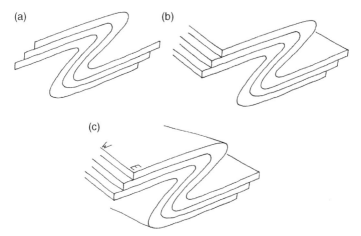

Figure 4.14 *Schematic drawing of a recumbent monoclinic fold. Equivalent to an echelon block, the three layers are displaced against each other to uncover their flat faces. (**a**) First, the frontal face is drawn, which is cut roughly perpendicular to the fold axis and represents a geographically vertical plane. (**b**) The edges of the layers are added to the frontal face. They run obliquely to the "left back" and mark the geographic horizontal. (**c**) The approximate orientation of the fold axis can be inferred if the layer edge is marked labeled with cardinal directions. The drawing does not, however, allow the precise dip angle of the fold axis to be determined. The top face of the upper layer and the bottom side of the lower layer are well visible with the present orientation of the drawing and can be filled with fabrics as an additional step.*

axes are represented only on the horizontal surface of the fold block and those with flat axes only on the vertical surface (Figure 4.16b). However, this form of representation has the disadvantage that it is not always clear which fold is older or younger.

4.2.4 Sedimentary Fabrics

Sedimentary rocks are characterized by a predominantly planar structure. Transitions between individual layers can be sharp or diffuse. Drawings of sedimentary rocks usually consist of stacked layers of different composition and structure (stratigraphic columns). Since linear fabrics are often missing, the drawings are mostly kept in 2D. Excellent examples of this, and (sometimes) instructions for drawing, are given in many books (Gwinner, 1971, Trewin, 2002, Coe, 2013, Prothero & Schwab, 1996).

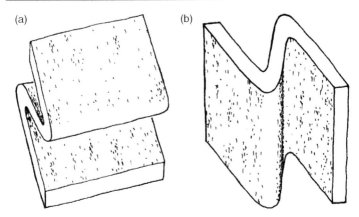

Figure 4.15 Schematic representation of differently oriented folds. *(a)* Orientation of fold axis perpendicular to the viewing direction. On the lateral face, the geometry of the fold appears strongly distorted. The fold crest is highlighted by contrast in brightness (different fillings by a mineral lineation). *(b)* Steep axis with fold crest facing the viewer. The bend is illustrated by different fillings and corresponding brightness contrast.

$\longrightarrow$

Figure 4.16 Schematic representation of folded folds and development of a drawing. *(a)* First, the outlines of an isoclinally and openly folded layer are sketched on a vertical and a horizontal plane. *(b)* In the second step, the—partially curved—fold crests are added and the outlines of the block are completed. This is the most demanding step, because the silhouette of the horizontal crest has to be drawn in two pieces that need to terminate exactly at the two bends generated by the vertical axes. Drawing the curvatures of the crest doesn't cause further difficulties. *(c)* In the third step, the margins of the crests have to be darkened, leaving the crests themselves blank, in order to mimic the curvature. The darkening is best simulated by short strokes representing a possibly existing lineation. But dotting would also do. In this stage, attention should already be paid to the correct appearance of shadows. In addition, the orientation of the strokes should support the image of curvature. In general, shadowing is applied to intensify the 3D appearance of the fold. *(d)* In the final step, shadowing can be (but does not need to be) intensified further to increase the 3D appearance of the drawing. *(e)* If one wishes to avoid the difficulty of drawing folded fold crests, folds with steep axes can be represented only on the horizontal surface of the fold block and folds with flat axes only on the vertical surface. However, in such a drawing, the chronology of folding remains ambiguous.

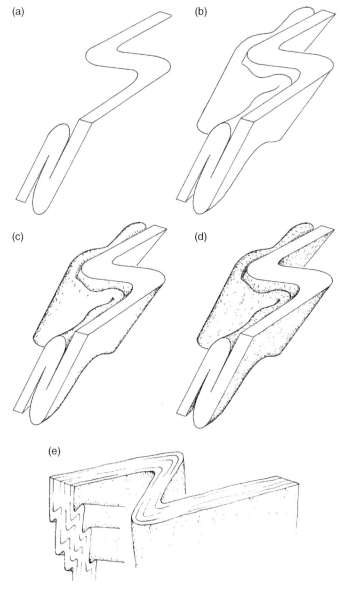

(a)

(b)

(c)

(d)

(e)

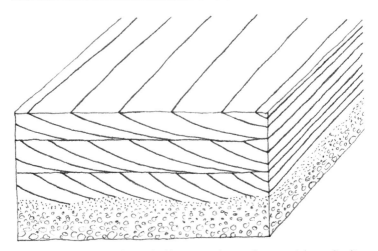

Figure 4.17 *Truncated cross bedding in sandstone above a grit layer. Grading in the grit layer is illustrated by dotting and various sizes of curls. The boundaries between cross bedding lamellae, mostly diffuse in nature, are schematized by solid lines. The 3D block allows the direction of the current to be estimated. Size of original sketch ca. A6; felt-tip pen with 0.1 mm line weight.*

Drawing sedimentary rocks in 3D is useful when linear fabrics are available. For example, flow directions can be derived from cross bedding structures or flute casts (Figure 4.17). Even the geometric form and extent of sedimentary bodies can be significant and are therefore worthy of being depicted in 3D. As is the case with metamorphic rocks, the blocks of sedimentary rocks should be bounded by orthogonal surfaces of which one is always oriented parallel to the bedding and the other is perpendicular to the bedding and perpendicular or parallel to the linears (if present). Because many sedimentary rocks are fine-grained, their structures are best represented through dottings of different types (Figure 3.4a,b and 4.6). This also helps to adequately represent gradations, filled desiccation cracks, flaser bedding, and so on. When sketching quickly, it often suffices to mark boundaries with lines.

4.2.5   Magmatic Fabrics

Mingling and mixing are distinctive processes during the often long periods of ascent and emplacement of magma in the continental crust (Gill, 2010). Moreover, magma is often deformed during its crystallization (Paterson *et al.*, 1989,

Vernon, 2000). This leads to structures that, in turn, provide information about the interaction between crystallization and deformation. Therefore, the effort of observing, measuring, and recording these structures in precise drawings is worthwhile. Contrary to metamorphic rocks, but similar to sedimentary rocks, the fabric inventory of magmatic rocks is quite limited, and the fabrics are often irregular and diffuse. For this to be drawn, increased effort is required.

Magmatic structures are also represented most easily in a block or echelon block. If the fabric is anisotropic, one block surface must be parallel to the planar or linear fabric (Figure 4.18a). In the next step, diffuse boundaries should not be sketched with solid lines, but rather pre-sketched with dotted ones. This facilitates the correct representation of shapes and spatial relationships of magmatic rocks as well as the, on occasion, diffuse or gradual transitions between them.

In a further step, the areas are filled with different internal fabrics (Figure 4.18b). Here, one uses the bigger crystals (if present) to visually elaborate flow textures, magmatic shear zones, accumulation of crystals, and so on. This also avoids the danger of filling surfaces too much and not having enough room to delineate different rock layers from one another through contrast. To save time, some areas may be left partially open.

In the final step—if necessary—the groundmass is drawn, dark inclusions are filled, and labels are added (Figure 4.18c). If samples were taken, the sample locations are marked in the drawing. The fabrics and internal structures in the drawing at hand contain significantly more information about the mineral content, granularity, orientation, and so on, than the label can convey.

Mingling and mixing fabrics can only partially be drawn strictly defined. Gradual transitions must be represented through differently dense dottings or dashes. This is quite possible with dots and short dashes (Figure 4.19), even under time pressure in the field (Figure 4.20). Light-dark contrasts of rocks are represented by different line densities and grain orientations by way of corresponding line orientations. Mixing domains with diffuse edges can be worked out with gradual changes in the density of the lines. While this requires an increased effort, it leads to drawings that reflect the natural fabric quite well and have a high information content. Even details of mixing structures like shape and the extent of schlieren, for example, can be represented in this way. The partial leaving open of areas of the drawing reduces both the drawing effort and time (Figure 4.21). The drawing is only filled with internal structures where important fabrics are present and only to the extent that the fabric characteristics become clear.

4.2.6 Complex Structures

The more structures a rock has, the harder it is to adequately represent them in a single block. The main reason is that planar or linear structures formed by successive

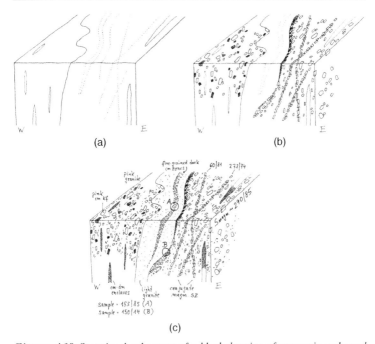

(a)

(b)

(c)

Figure 4.18 *Stepwise development of a block drawing of magmatic rocks and fabrics. (a) Orthogonal block with two surfaces perpendicular to the magmatic layering and one surface parallel to the magmatic foliation. Diffuse structures and boundaries are marked by dotted lines, sharp boundaries by solid lines. (b) Feldspar and biotite phenocrysts (open and closed rectangles of variable sizes) illustrate the orientation of magmatic flow planes and shear zones. (c) Additional filling, with short strokes (biotite platelets), highlights the variably strong flow alignment in different rocks: random in fine-grained mafic enclaves (slim, dark lenses), weak in light granite (mimicked by short strokes and dotting), and considerable in porphyritic ("pink") granite and conjugate magmatic shear zones. The variably intense filling mimics the natural contrast in brightness of the rocks. In addition to labeling, the orientations of magmatic planes and of two samples, A and B, are specified. The approximate positions of the samples relative to the magmatic structures are given. Late-Variscan granitoids at Punta di Vallitone (west-coast of Corsica, France); after a field drawing; outcrop KR4795 (field book 40, Kruhl, 2005); size of the original drawing ca. A6; felt-tip pen with 0.2 and 0.3 mm line weight.*

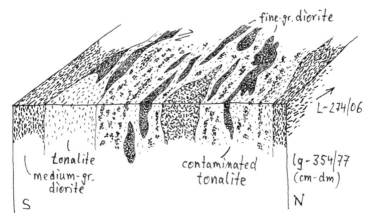

Figure 4.19 *Block drawing of tonalite and diorite injections into an active late-Variscan shear zone. Shearing caused different mingling, mixing, and deformation structures. Grain sizes, compositions, and flow patterns are represented solely by biotite and amphibole platelets (variably long and thick strokes with locally differently dense distribution). Elongate layers, that is, layers more strongly deformed during crystallization, show tight amphibole and biotite alignment, whereas more weakly deformed regions exhibit an accordingly weaker alignment or no alignment at all. Diorite-tonalite-granite sequence at Abbartello (Golfo de Valinco, Corsica, France); re-drawn field drawing; outcrop KR4846 (field book 40, Kruhl, 2005); size of original drawing ca. A6; felt-tip pen with 0.2 and 0.3 mm line weight.*

and independent processes are usually not parallel to each other. Thus, they cannot be pictured simultaneously on one or even a few surfaces.

This means, structures must be drawn stepwise and then put together. This way, the drawing develops in the same manner as the observation progresses. Nevertheless, one should still observe first and recognize as many structures as possible before beginning the drawing process. This goes for relatively simple echelon blocks as well as for complex structures.

Principally, the simple and large structures, like the layers of an echelon block, are drawn first. They form the backbone of the drawing, which will be expanded upon in the next steps (Figure 4.22a). This framework consists of simple parts—mostly plates that represent individual layers or folds. It is important that these parts are kept open, meaning they are only partially bounded, and that they don't take up the entire available drawing area. At least two sides should have

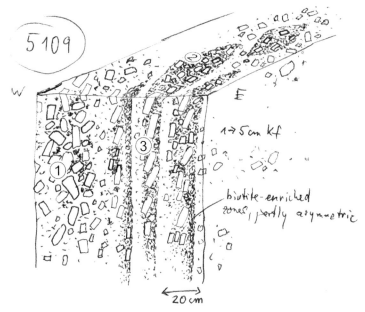

Figure 4.20 *Block drawing of a syntectonically crystallized granite, with diffuse accumulations of K-feldspar megacrysts (1), biotite schlieren and lenses (2), and shear fabrics (Kf-tiling) (3). One of the three orthogonal block faces is cut parallel to the magmatic flow plane. The other two faces are oriented perpendicular to the flow plane. From these two faces, a roughly vertical magmatic lineation can be inferred based on the different shapes of dark lenses and K-feldspar sections. Boundaries of layers are marked only locally by broken lines but are generally indicated by differences in line weight and density of strokes. Dark schlieren and lenses with unoriented small biotite platelets (short strokes) represent mingled mafic magma, while small spots of randomly oriented biotite indicate an advanced stage of magma mixing. In order to avoid an overfilling of the drawing and poor visibility of magmatic structures, the fine-grained groundmass is not drawn. Labeling is limited to the essential. Road cut between Casar de Cáceres and Arroyo de la Luz (Cabeza de Araya Batholith, Spain); outcrop KR5109 (field book 42, Kruhl, 2011); size of original drawing ca. A6; black ballpoint pen; expenditure of time for observation and drawing ca. 20 minutes.*

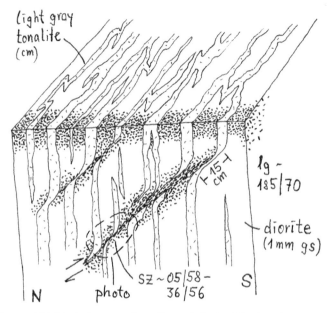

Figure 4.21 *Block drawing of tonalite and diorite injections into an active late-Variscan shear zone with two faces perpendicular and one face parallel to the magmatic layering. The contrast between dark diorite and light tonalite is generated by differently dense dashes of varying line weights. The drawing is not to scale, that is, there is no generally applicable scale. However, scales of different structures can be given: cm-thickness of the tonalite layers, 1 mm for grain size of diorite, and the displacement at the magmatic shear zone. Layering (lg) and shear zone (sz) are the only structures with measurable orientation. The two values of the shear zone define the range of orientation of several shear zones throughout the outcrop. The drawing provides more information than given by labeling. The parallel layer boundaries on the frontal face and the irregular, "wrinkled" boundaries on the horizontal top face point to anisotropic injection of the tonalite or sub-horizontal flow of the mingled magmas. The generally unoriented mafic mineral grains (short strokes) of diorite and tonalite indicate absence of differential stress during magma crystallization. Only the mafic mineral grains of the diorite are aligned in both northward inclined shear zones. Consequently, during shearing, diorite existed as melt-crystal mush while tonalite existed predominantly as melt. Diorite-tonalite-granite sequence at Abbartello (Golfo de Valinco, Corsica, France); re-drawn field book sketch; outcrop KR4843 (field book 40, Kruhl, 2005); size of original drawing ca. A6; felt-tip pen with 0.2 and 0.3 mm line weight.*

143

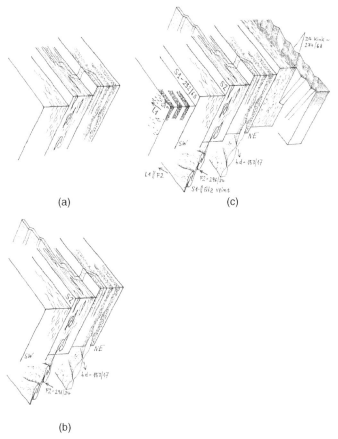

(a)

(c)

(b)

Figure 4.22 *Stepwise development of an echelon block drawing with differently oriented structures.* **(a)** *Similar to a "normal" echelon block, different rock layers are displaced against each other and partly filled with internal structures. At this early phase of drawing, the layer boundaries remain partially incomplete, and the flat faces of the layers, facing the viewer, remain blank.* **(b)** *In a second step, single layers (dykes, boudinaged layers) are drawn backward or downward away from the block to illustrate fold axes, linears, and so on. In this stage of drawing, structures can already be labeled.* **(c)** *Newly observed rocks are added as plates to the front or back of the echelon block. The schist layer with kink bands is extended outward to illustrate the orientation of the kinks. Further labeling is added.*

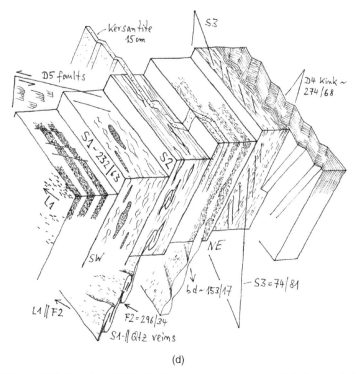

(d)

Figure 4.22 *(continued)* ***(d)*** *In the final step, young structures (foliations, faults) are drawn and final additions are made. In order to avoid corrections, these structures are placed in gaps of the drawing, if possible. Single planes are drawn outward to better illustrate their orientation. Size of the original drawing ca. A5; felt-tip pen with 0.2 mm line weight, in part with obliquely set lead.*

ample space for further "attachments." The individual plates can already be filled with structures early on to differentiate between various rocks.

There are two basic types of "attaching": (1) drawing existing structures separately to make them (more) visible, and (2) the attachment of new structures to the existing drawing. Drawing structures separately is always useful if the geometries of layers or structures are hidden in the block. It is easiest to remove the structures from the block just enough to make their form or extent visible (Figure 4.23). This is useful for rod shapes (boudins, pebbles, or crystals), for example, that lie on the edge of the block, but that are—cut from a block surface—not completely

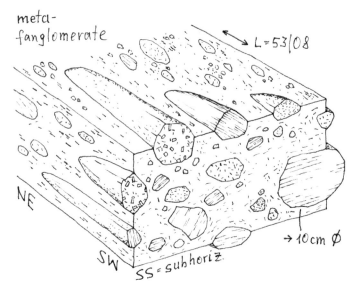

Figure 4.23 *Schematic block drawing of a weakly foliated but intensely stretched metafanglomerate. The orientation of bedding is inferred from the outcrop and indicated, although not visible, in the drawing. The initially round or slightly oval pebbles of metapsammite and volcanic rocks are deformed and strongly stretched. Some rods are partially lifted out of the block for better representation of their shape and orientation. This emergence should already be considered when drawing the block outlines. In addition to dottings, lineations on the rod surfaces (short strokes) and intersection lines with foliation planes are used for shadowing and, consequently, increasing the 3D appearance of the drawing. In cross section, the metapsammite is illustrated by parallel, broken lines representing the schistosity, and the volcanic rocks by unoriented small rectangles and short strokes, representing feldspar and biotite, respectively. Metafanglomerate of the south-east Sardinian Paleozoic basement (Gerrei). Re-drawn field sketch; outcrop KR5118 (field book 42, Kruhl, 2012); size of original drawing ca. A6; felt-tip pen with 0.1 mm line weight.*

recognizable in 3D. It is sufficient to highlight only some of the structures in this way.

Highlighting those structures whose spatial orientation is otherwise non-recognizable, is even more important. This way, the orientation of fold axes can be made visible (Figures 4.22b and 4.24). One also gains space on which structures (e.g., lineations) can be represented. If all these structures would make up individual layers of the echelon block, this would lead to significantly larger and more compact drawings, which take up more space. Not least, structures in the existing block can be highlighted and attached to the block in almost any direction. This considerably increases the flexibility of the drawing's design. If it is foreseeable that the drawing will be labeled in detail, it is appropriate at this stage to name individual structures and assign them measured values.

If one keeps the surface facing the viewer at an angle open and far enough away from the edge of the field book, additional layers can easily be added (Fig. 4.22c). It usually (not always!) doesn't matter whether these layers are also next to each other in the outcrop. The attachment of layers is important when it would otherwise require too many corrections to add new structures to the drawing. Moreover, it is significant that there is space next to the surface facing away from the viewer. It is here that larger structures can be inserted without disturbing the rest of the drawing.

Even when a drawing is almost completely closed, there are usually enough gaps and spaces surrounding the drawing to draw individual planes (foliations, fault planes, etc.; Figure 4.22d). This way, their orientations can be clarified and structures can be marked on them. It does not matter if foliation planes, for example, cannot exist in isolated form and therefore these planes exist here only virtually. The intention of the drawing is not to show reality but a model of reality that includes all geologically relevant information and depicts it effectively.

In the late stage of drawing and observing, it often proves to be useful when the layers are only partially filled with internal structures. This way, structures observed later can still be inserted into the drawing without significant corrections. If the fill is heavy, personal skill is required in order to insert new structures or adapt them to old ones without ugly exaggerations. The open borders of the individual layers can be closed, but do not have to be. In case the drawing is excessively filled or there is no space for any additions, the option of drawing up a second echelon block for the other structures remains.

Drawings of complex folds develop in a similar way. First, a simple base block, which adapts to the main fold structure but can already contain internal fold fabric, is drawn (Figure 4.25a). Again, it is convenient to omit parts of the block's boundaries. In the next step, more folds, belonging to the younger generation and folding the older structures, can be added to the back of the block (Figure 4.25b).

Spatial separation of different fold generations helps avoid time-consuming drawing. If other structures that can no longer be inserted into the existing drawing without major corrections are discovered, like a younger dyke that breaks through the folds, for example, an extension of the block can be a useful remedy. The structures must not be attached directly to the existing block. They can be free standing with a gap in between. The important thing is that folds of the same generation, for example, have the same geometry and that their axes and axial planes have the

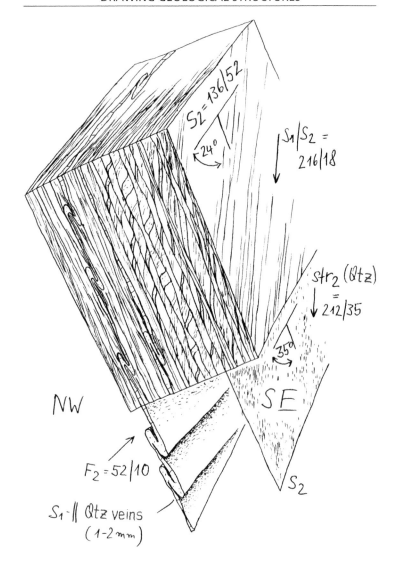

$S_2 = 136/52$

$24°$

$S_1/S_2 = 216/18$

$str_2 (Qtz) = 212/35$

$35°$

NW

SE

S_2

$F_2 = 52/10$

S_1-∥ Qtz veins (1-2 mm)

same orientation. It does not matter if structures in one half of the drawing do not continue in the other half, for the drawing represents the geological structures only schematically and in the greatest possible clarity and simplicity.

Even in this quite complex sketch, new structures can be incorporated in gaps without a problem (Figure 4.25c). In rocks that have been deformed multiple times, the foliation planes are of particular importance. If they are precisely sketched with their mutual transective relationships, the sequence of deformations can be made absolutely clear. These planes are also used to enhance the 3D picture of the drawing. If younger structures, like the S2-parallel quartz veins in Figure 4.25c, for example, need to be inserted into older ones, narrowings in the structure can be used to avoid overlaps of lines.

In the final step, detail structures are placed on different surfaces (especially lineations), prominent internal structures are added (porphyritic vein structures and the cross fractures typical of quartz veins (Chap. 3.3—Figure 3.4m)), and labels with measured values are included (Figure 4.25d). When dealing with recognized structures, it is often easier to measure them, incorporate them in the drawing, and label them all in one go. These early labels can, however, hinder the development of the drawing and must sometimes be "circumvented." The finished drawing need not be completely bounded and may remain incomplete—even more so than in this example (Figure 4.25d). It is sufficient if the structures are only locally implied. The information content is not decreased; it is just the drawing that becomes clearer.

Figure 4.24 *Schematic block drawing of a strongly foliated metapsammopelite, with characteristic fabric inventory. A folded first schistosity, S1, is preserved as relics in quartz-rich layers. The second schistosity, S2, in a quartz-rich layer forms the lateral face of the block. The T-sign marks the strike and dip directions on the S2 plane, the intersection linear, S1/S2, is shown. This schistosity is drawn out of the block as a single plane, in order to visualize the stretching lineation, str2. In addition to the orientations of the stretching and intersection lineations, their angles versus the horizontal are given. Initially S1-parallel, mm-thick quartz veins are characterized by typical cross fractures. The dotting indicates the small grain size of the veins and, therefore, the recrystallization of quartz. The orientation of the axes, F2, of the isoclinally folded quartz veins is depicted by a quartz vein drawn downward away from the block. Str2 on the surface of this vein is illustrated by stretched quartz grains (short strokes parallel to F2) and also used for increasing the 3D appearance of the drawing. The quartz vein is transected by the lower block boundary—an indication that the outer part of the vein was added after completion of the block outlines.*

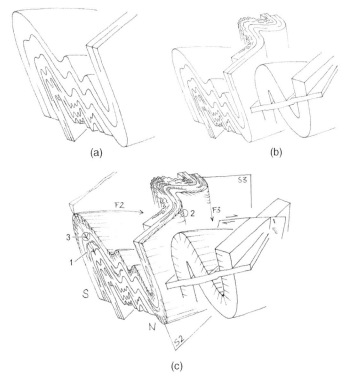

(a)   (b)

(c)

Figure 4.25 *Stepwise development of a drawing of multi-fold folding.* **(a)** *Single-folded layers form the basic module of the drawing. The upper flat cut surface of the layers marks the horizontal. Accordingly, the fold axes show a flat to moderate dip. The drawing is kept open toward the back, in order to facilitate the annexation of further structures.* **(b)** *A late, steep fold, together with small secondary folds, is added to the back of the block. Thus, one avoids the difficulty of drawing of folded fold bends. A second fold of the early fold generation, transected by a young sub-horizontal dyke, is placed to the right of the block. This prevents the insertion of the dyke into the block, usually associated with complex corrections. The second fold, too, is kept open towards the back.* **(c)** *In the next step, detail structures are inserted into the drawing, specifically foliation planes of the three deformation events, S1-parallel quartz veins (arrow 1), and a late fault. Individual planes are drawn out of the block in order to make lineations on these planes visible. The schistosities are also important for emphasizing the chronology of deformation*

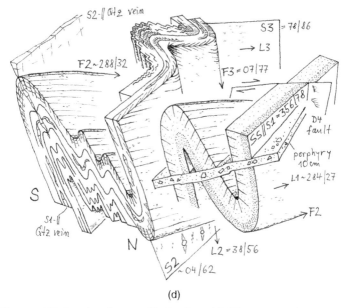

(d)

Figure 4.25 (continued) events. The bending of S2/SS intersection lines around an F3-crest (circle 2) supports the chronology of folding. The neckings of an S2-parallel quartz vein are positioned in such a way that the vein is optically not transected by already sketched (older) layer boundaries (circle 3). **(d)** Subsequently, layers are filled, lineations are placed on the different foliation planes, and measurements and labeling are plotted in the drawing. The internal fabrics are also used for shadowing to increase the 3D appearance of the drawing. Size of original drawing ca. A5, felt-tip pen with 0.2 mm line weight and lighter due to obliquely set lead.

As is the case when drawing thin sections (Chap. 2, Figure 2.8c, 2.20b), the enlargement of parts of sample and outcrop drawings is a good way of connecting structures with different scales (Figure 4.26). It makes it possible to draw structures on a large scale without having to fill them with too many details. This saves time and prevents the drawing from becoming visually overloaded. Fillings only appear where they are meant to support large structures and in areas that are to be enlarged. The shapes of these areas do not matter. Usually they represent circles

151

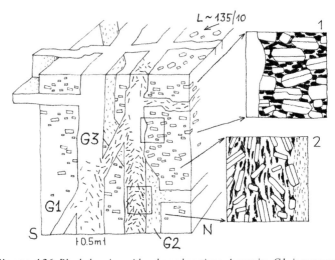

Figure 4.26 *Block drawing with enlarged sections. A granite, G1, is transected by dykes of two later granite generations, G2 and G3. The sub-horizontal magmatic foliation of the early granite, G1, is represented by tabular feldspars and is illustrated only locally. Within the dykes of the two late granites, the flat faces of the euhedral feldspars are oriented approximately parallel to the dyke walls or form cross arcs. The large thin-tabular feldspars in granite, G2, are represented by long strokes, the smaller feldspars in granite, G3, by short, slightly thinner strokes. Mostly feldspars mark the magmatic foliation. Therefore, the groundmass of the rocks does not need to be illustrated. This saves much time and keeps the drawing "light." The enlarged sections clarify that the represented rocks (usually) appear lighter in the drawing than in reality. This discrepancy to nature is fully compensated by gains in time and clarity of the drawing. The enlarged section 1 of G1 represents the grain fabric. It illustrates the high amount of coarse, aligned biotite in the groundmass and also that feldspar crystals are cut by dykes of granite, G3. The G3 area is kept blank, because these fabrics are already shown in section 2. This enlarged section covers the border region between G2 and G3. Again, it clarifies the high amount of biotite in G3 and represents details of the orientation fabric in G2 and G3. The scale in the principal drawing is only related to the thickness of the dykes. Consequently, three differently sized ranges are present in the total drawing: the m-dm-range for the dyke system, the cm-mm-range for the feldspar phenocrysts, and the mm-μm-range for the two enlarged sections. Re-drawn, merged field sketches of syntectonic, late-Proterozoic granitoids (Porto Belo, Brazil); outcrops KR4226 and KR4230 (field book 36, Kruhl, 1996); size of original drawing ca. A5; felt-tip pen with 0.1 and 0.2 mm line weight.*

or rectangles/squares. The enlarged sections are placed next to the drawing and connected with lines to their place of origin, or marked with numbers or letters.

If a drawing already contains numerous details, it may be worthwhile to depict small-scale structures in additional drawings (Figure 4.27). This is especially

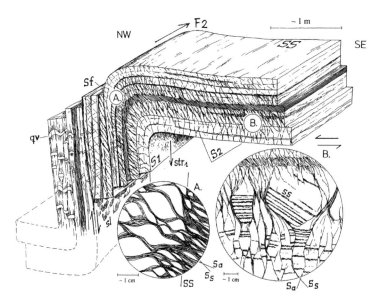

Figure 4.27 *Schematic drawing of an m-scaled fold in Devonian metapsam-mopelitic layers ("Spitznack" fold) of the Rhenish Massif (Middle-Rhine region near the Loreley, Germany). See also photograph and sketch in Figure 1.3. The fold resulted from top-to-the-northwest movement. The long and short limbs of the fold are mainly filled with foliation planes, S2, of the second deformation event. Special attention was paid to precisely illustrate the variable orientation of the two sets of S2 planes that developed in pelitic and psammitic layers. The rotation of small fragments of pelitic and psammitic bedding, SS, between these two sets, Sa and Ss, is clarified in the two enlarged sections, A and B. Additionally, structures important for the interpretation of the deformation history of the rock, such as the stretching lineation (str1), bedding-parallel shear planes (sf), slickensides (sl), and sheared and horizontally compressed quartz veins (qv), are represented. Modified after Zurru and Kruhl (2000, figure 33); size of original drawing ca. A3.*

relevant if these small-scale structures bear important information that cannot be obtained from the main drawing. In such a case, it can even be justified to make the additional drawings nearly as large as the main drawing.

If one uses the opportunity to already represent specimen or outcrop drawings at different scales, the to-be-enlarged areas may already lie in the cm-mm-scale. This means, a further enlargement in the outcrop drawing will present the structures in the grain scale, that is, in the thin-section range, without a problem. The entire drawing can theoretically represent fabrics from the dm- to m-scale down to the micrometer range.

If an outcrop contains too many different structures, like a mix of sedimentary and deformation structures, for example, it is worthwhile to take a larger sample and carefully draw these structures at a large scale and with precision (Figure 4.28). Again, these areas need not be completely filled. The high number of different structures in this drawing, however, requires almost complete execution of the drawing. The effort is worthwhile. Beyond the frugal labeling, almost every detail structure offers important information about the sedimentation and subsequent metamorphic and deformation history of the rock. Of course the structures need to be readable in order for the drawing to be used as a data archive. If the drawing is meant for more than just "household use," more detailed labeling should be added.

$\longrightarrow$

Figure 4.28 *Folded and foliated psammopelitic layers from the Tertiary Flysch of the Central Alps (quarry Gasparini near Seedorf, Uri, Switzerland); sample KR3756. The sedimentary fabrics comprise (i) irregularly thick, partly wedging, quartz-rich bedding (dotted layers), (ii) fine cross bedding lamellae, and (iii) grading in pelitic layers. The deformation fabrics mainly consist of (i) early-D1 extension quartz veins in psammitic layers, (ii) a first bedding-parallel schistosity, S1, and stretching lineation, str1, in pelitic layers, (iii) S1-parallel quartz veins, (iv) late-D1 shear planes, (v) a second schistosity, S2, and second stretching lineation, str2*, in pelitic layers, (vi) a second folding, F2, with extension quartz veins in fold cores, and (vii) S2-parallel quartz veins. The sparse filling of the psammitic layers with internal fabrics increases the contrast to the pelitic layers and accentuates the strong variation of the layer thicknesses. To save time, the pelitic layers are only partially filled. The drawing clarifies that the rock layers showed ductile but no crystal-plastic behavior during deformation. Ink drawing; original size ca. A3.*

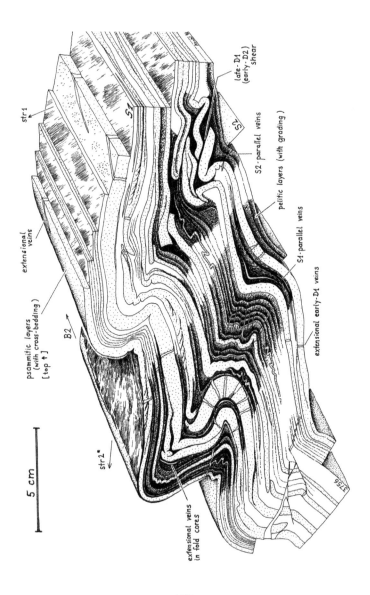

4.3 Field Drawing

In the previous chapters, the stepwise and systematic evolution of drawings was handled in detail. This was meant to clarify the principles of drawing. In the field, however, the environment is very different from a table in a warm room. Just as we observe and measure stepwise and often in jumps—with mistakes and disruptions—the drawing in the field develops in jumps and with mistakes and subsequent corrections. If the fingers become stiff with cold, the wind dishevels with pages of the field book, we stand with cramped muscles on a steep slope, and time is limited, then a "nice" drawing is the last thing that crosses our mind. Instead precision, speed, simplicity, and conciseness stand in the foreground. Therefore, it should be stressed again here: Clean and aesthetic drawings can bring much joy to the viewer and are not out of place in presentations and publications. Ultimately, there are only two things that are important in the field: first, that we watch closely, observe intensely, and, that through our observations, gain knowledge about geological structures and processes; and secondly, that we bring our observations to paper in a concise drawing language, and our field book becomes an information-rich and easy-to-read data archive (Figures 4.29 and 4.30).

This means, a field drawing is usually created as quickly and as economically as possible. Even if there is nothing more than a single layered and foliated rock (Figure 4.31), or layers of two different rocks, with a main foliation, a lineation, and a few quartz veins and boudins (Figure 4.32), visible in an outcrop, it is still worth making a drawing. Not only is it faster than writing a descriptive text, it also depicts the structures more clearly. In addition, it can only be emphasized time and

---

$\longrightarrow$

Figure 4.29 *Scans of two field book double pages with drawings from one outcrop in the Finero Magmatic Complex (Ivrea Zone, Valle Cannobina, northern Italy). The various rocks and their fabric inventory are shown in five similarly oriented drawings, reflecting the chronology of observation. After the first block, with the main layering and lineation, was generated and labeled, new structures were observed in the neighboring layers and represented stepwise in new drawings. Drawing fabrics in separate blocks simplifies the representation of single structures and increases the clarity of the illustration. The layering of the metagabbro, which consists of different amounts of plagioclase, pyroxene, amphibole and garnet, is shown in two of the five blocks and only weakly indicated in the other three. Each structure also appears in only one block and only one measurement, or range of values, is related to it. The orientations of samples and their structural inventory are given, and the approximate sample positions are marked in the drawings. In general, the drawings are kept incomplete to save time. The thickness of different types of layers is given. However, other structures are not to scale. Field book 35 (Kruhl, 1996); size of a double page ca. A4; labeling in German; black felt-tip pen.*

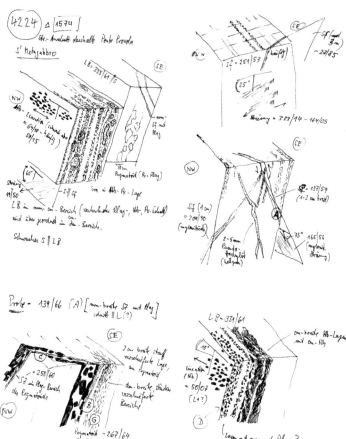

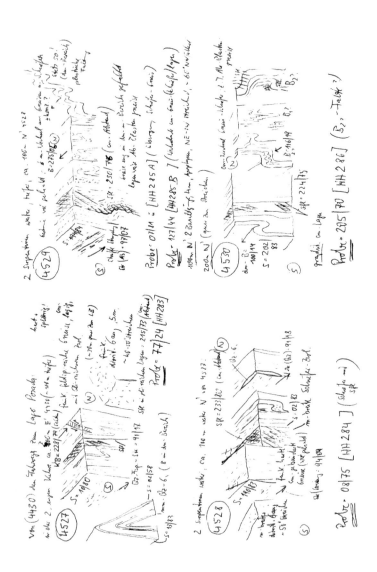

time again that, in contrast to textual information, graphic information in a field notebook can quickly be recognized and processed.

The following also goes for fast drawings in the field: The drawing should be oriented so that important surfaces are clearly visible and 3D structures (ex. folds) are easy to represent (Figures 4.33, 4.34, and 4.35). The drawing is usually only partially bounded. Under circumstances, it can be "cut off" from the edges of the field book. Much of the labels and values may be added early on to the surfaces on which they belong. Only then are areas with internal fabrics filled. This way, the labels can be better allocated than if they were listed next to the drawing. Even when one wants (or has) to draw quickly, one should take the time to emphasize the 3D effect of the structures through meaningfully placed light-dark contrasts. This increases the readability of the drawing.

Especially when observing and drawing in the field, two processes that are often approached unsystematically or cannot be approached systematically at all, the strengths of a semi-schematic sign language can be applied: the speed of schematization, the high recognition value or symbolic structures, and the flexibility of a free-form drawing, which is not tied to the spatial relationship of rocks and structures or their scales. There are (almost) no limits in free-form drawing. If time is short, the drawing is filled with only enough internal structures as necessary. If time as well as peace and quiet are available, one can treat oneself to the pleasures

◄───

Figure 4.30 *Scan of a field book double page with drawings from the Variscan basement of the Baronie (eastern Sardinia, Italy), mostly consisting of schists and different types of gneisses. The fabric inventory of four different outcrops along a road, cut perpendicular to a large-scale fault (Posada Asinara Line), is sketched and labeled. Drawings and labeling are placed too closely, compared to those in Figure 4.29, and would not allow the blocks to be extended in case of new observations. All blocks are equally oriented so that the direction of view remains constant. Thus, similarities as well as differences between structures can be detected quickly. For example, the lineation on the main schistosity always has the same lineation (from block to block). The same is true for a set of late shear planes, ssl, which appears in all mica-rich layers. Differences are obvious with respect to the folding of the main schistosity. The geometry of the folds, as well as the sense of folding, changes along the profile. Late mafic veins, not affected by deformation, are drawn separately. Thereby, complicated insertion of the veins into the already present blocks is avoided. The orientations of samples and their structural inventory, together with the thicknesses of different types of layers, are given. However, other structures are not to scale. Field book 37 (Kruhl, 1999); size of a double page ca. A4; labeling in German; black ballpoint pen.*

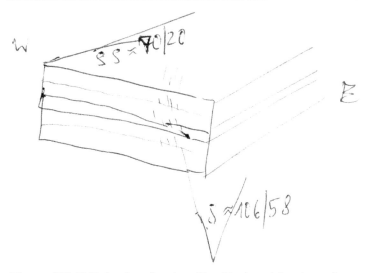

Figure 4.31 Field drawing of gneiss with subhorizontal layering and steep schistosity; Moinian of the Scottish Highlands west of the Great Glen Fault. This drawing, made in a few seconds, violates "drawing rules." It still, however, serves to quickly document the sparse rock fabrics and measurements. Thus, precise drawing and a scale are not required. Sample KR1754 (field book 18, Kruhl, 1979); original size of the drawing ca. 8 x 5 cm; ballpoint pen.

of developing and making the drawing "presentable." If one discovers new structures that can no longer be integrated or added into the existing drawing, a new drawing should be made next to or below the first (Figure 4.36). The new drawing should be oriented the same as the old one and adopt the same (local) scales and symbols.

It is also not the end of the world to infringe upon the "proper rules." When drawing quickly, the lines along the block corners often remain unconnected, the layer or foliation planes are irregular and not strictly parallel, and existing borders or whole parts of the drawing are often drawn over when structures are added or things need to be corrected. Precision and cleanliness of the drawing are not nearly as important, as long as the geological message is clear.

If one has the time in the field and the rock structures are varied and complicated, it is definitely worthwhile to start the drawing after a long period of observation and to compose a drawing that includes all the structures (Figure 4.37). This helps the coherence of the structures and often leads to a drawing that is admittedly

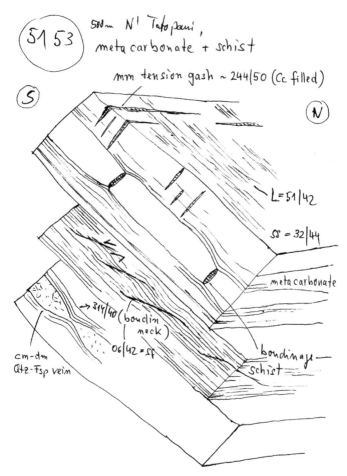

Figure 4.32 *Field drawing of low-metamorphic, foliated, and boudinaged metacarbonate and schist layers (Lower Greater Himalayan Sequence), with only a few distinctive structures. The drawing is kept accordingly sparse. Lower layers and part of the internal fabric were added after labeling. A scale is given by the thicknesses of quartz-feldspar veins and tension gashes. Outcrop KR5153, Kali Gandaki Valley, north of Tatopani (Baglung, Nepal); field book 43 (Kruhl, 2013); size of original drawing ca. B6; ballpoint pen.*

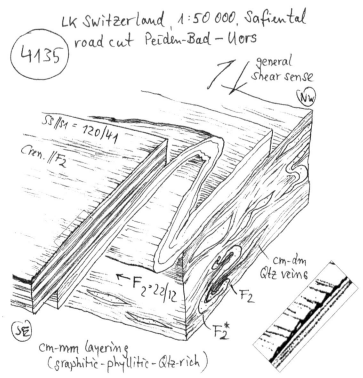

Figure 4.33 *Foliated and folded phyllitic layers with S1-parallel quartz veins, stretched to lenses and locally isoclinally folded. Recrystallization of these veins is marked by light dotting. The D2 crenulation of the layering-parallel schistosity was added relatively late and small corrections of the initially straight block edges had to be made (shown in the detail sketch). Lineations are used to increase the 3D appearance of the fold block. Labeling is limited to the necessary. Because fold axes and lineations are parallel, no additional measurements are plotted. Instead, other information (e.g., "general shear sense") is added to the drawing. An approximate scale is given by the thicknesses of layers and quartz veins. Field drawing; outcrop KR4135; field book 35 (Kruhl, 1994); Safiental, Switzerland; size of original drawing ca. A6; ballpoint pen; original German labeling replaced by English.*

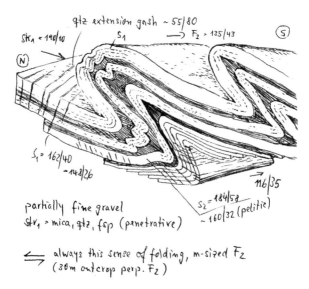

Figure 4.34 *Field drawing of double foliated and folded psammopelitic layers. Stretching and intersection lineations serve to increase the 3D appearance of the drawing. The strong second foliation of the pelitic layers is highlighted. The psammitic layers are only weakly foliated by the first and second schistosities, S1 and S2, and appropriately sparsely filled, in agreement with the brightness contrast in nature. Thus, details of folding are accentuated. The geometry of folds, with thickened fold crests, thinned limbs, and pile and fan position of S2 planes, points to ductile deformation and the strong contrast in strength between psammitic and pelitic layers. Because the foliation planes show variable orientations, the measured values are related to specific locations. The general shear sense of the D2 folding is inferred from the total outcrop. The drawing is not to scale and, therefore, cannot include a scale. Instead, the approximate thicknesses of layers and sizes of folds are indicated. Safiental, Switzerland; outcrop KR4137 (field book 35, Kruhl, 1994); size of original drawing ca. A6; black ballpoint pen; original German labeling replaced by English.*

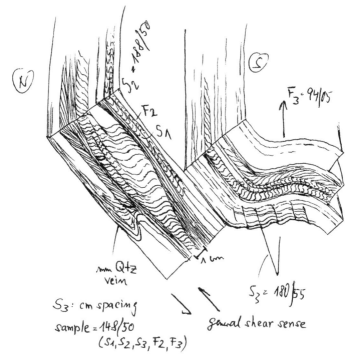

Figure 4.35 *Field drawing of folded and foliated, cm-thick metapsammopelitic layers (Permo-Triassic cover of the eastern Gotthard Massif near Ilanz, Switzerland). The drawing serves to clarify the three types of foliation and folding, and their chronology. Specific care was taken to illustrate the variation of crenulation in relation to the different mica content of the layers. The orientation of the sample, as well as which structures are visible at the sample, is recorded. Outcrop KR4136 (field book 35, Kruhl, 1994); size of original drawing ca. A6; ballpoint pen; original German labeling replaced by English.*

Figure 4.36 *Folded and foliated metapelite (Permo-Triassic cover of the eastern Gotthard Massif near Ilanz, Switzerland). Fillings and labeling are kept as sparse as possible in this typical "impure" field drawing. Foliation planes are largely drawn around quartz-carbonate veins, in order to increase the contrast. Little value is attached to graphical precision; more, however, to the relationship between the structures: (i) parallelism and discordance of quartz-carbonate veins*

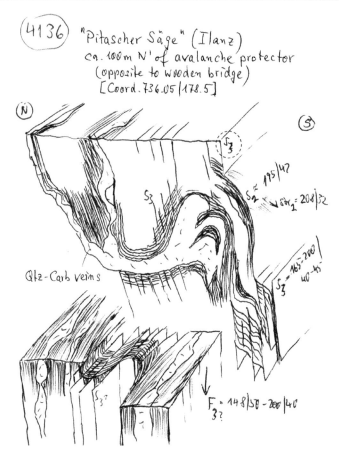

Figure 4.36 *(continued) to S2, indicating the variable age of vein formation in relation to foliation, (ii) folding of S2 between S3 planes, and (iii) the locally steep, as well as flat, F3. Retrospectively, S2 was categorized as S3 (correction within the broken circle). The range of strike and dip values of S3 and F3 reflect the spatial variation of these structures. It is difficult to show the folding of S2 planes between S3 planes in the upper (first) drawing which is mostly a vertical cut. Therefore, a second drawing with dominant horizontal top face is placed below the first one, as far as space allows. Field drawing; outcrop KR4136 (field book 35, Kruhl, 1994); size of original drawing ca. A5; ballpoint pen; original German labeling replaced by English.*

165

complicated but by all means clear and very instructive. Even in such a large-scale drawing, the structures are not necessarily represented exactly how they appear in the rock. This usually leads to accumulations of structures that make the picture very unclear. Even in such a drawing, it is useful to leave gaps in order to save time and to make the drawing easier and more manageable.

If larger structures are clearly developed in an outcrop, it makes sense to first construct a relatively simple overview drawing of these structures (Fig. 4.38). This drawing can of course already include smaller structures and labels. Further details will, however, be outsourced in additional smaller-scale drawings whose locations are marked in the overview drawing. This way, on the one hand, the larger structures are linked to the smaller ones and, on the other hand, an overload and "unreadability" of the drawing is avoided.

Figure 4.37 Field drawing of foliated and folded metapsammopelites. The drawing developed from left to right as two blocks side by side. Subsequently, folds are attached to the front and to the right of both blocks, in order to illustrate the orientation of the steep fold axes. If the attachment is shifted downward, as in the right side of the drawing, part of the lateral block face is kept free and allows structures (the sub-horizontal L2, in this case) on this face to remain visible. For clarity, structures are drawn only locally. This reflects reality as well, because different structures are related to different mica content of rock layers and, accordingly, appear locally. Due to their variable mica content, metapsammopelites excellently depict different deformation structures. Relics of sedimentary structures, for example, fine cross bedding (1), can be recognized occasionally in psammitic quartz-rich layers, as well as the first schistosity, S1, (2) and first lineation, L1, (3). During a later stage of drawing, S1-parallel quartz-vein rods (4) are attached to the lateral block faces, in order to illustrate their steep orientation and the related lineation. The lineation is also used to clarify the 3D shape of the rods. S1- and S2-parallel quartz veins are dotted to indicate quartz recrystallization and are outfitted with the characteristic cross fractures (5). In order to illustrate the spatial orientation of later structures, these structures are either drawn laterally (S3 schistosity [6]), upward out of the block (late quartz veins with pseudo folds [7]), or represented on two orthogonal block faces (kink bands [8]). The horizontal scale bar is related to the thickness of the layers. Cape Conrad coastal profile (Mallacoota, Victoria, Australia); outcrop KR5032 (field book 41, Kruhl, 2009); size of original drawing = double page of the field book, that is, roughly A4; ballpoint pen.

166

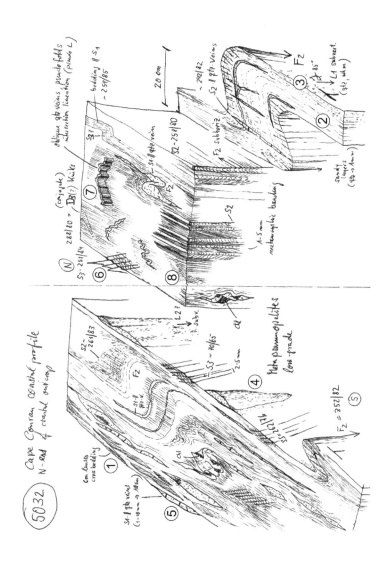

(5032) Cape Conran Coastal profile
N-end of coastal outcrop

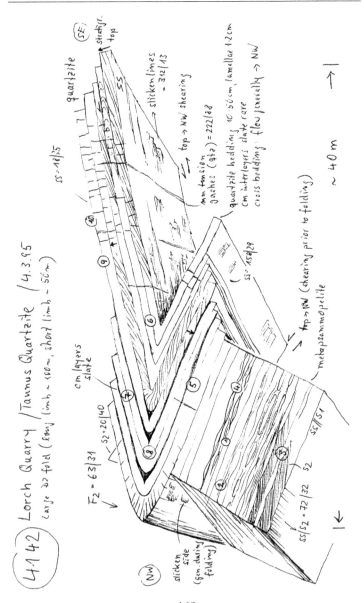

4.1.42 Lorch Quarry / Taunus Quartzite / 4.3.95
Large D2 fold (long limb ~ 100m, short limb ~ 50m)

168

4.4 Digital Processing

What applies to thin sections is even more true for the production of sample and field drawings. Apart from technical difficulties in the field, drawing using primarily the computer comes mostly with disadvantages. It is cumbersome, slower, and more inaccurate when it comes to details. Internal fabrics are represented schematically at best and fine structures are usually left out completely.

When it comes to drawings for presentations or publications, which depend on cleanliness and clarity, the digital processing of drawings has many benefits. Many of the typical weakness or mistakes that often occur during quick field drawings can be resolved: (i) amplification or reduction of too weak or too strong contrasts, respectively, (ii) elimination of stains, (iii) removal of "wrongly drawn" planes, and (iv) elimination of structures that are redundant or fill the drawing too much.

In addition, labeling can be improved and updated. During rapid field drawing, manual labels may be unreadable to later viewers unfamiliar with the handwriting. Computer fonts in different sizes and line weights can remedy this situation, especially if the drawing is complex and requires detailed labeling (Figure 4.39a). Even backing the label with a white background can make it more readable and allows the label to be put directly in areas filled with internal structures, for example.

Figure 4.38 Synoptic field drawing of sequences of folded Variscan quartzite and metapsammopelite; ca. 30 m long, NW-SE oriented quarry face in Taunus Quartzite (Lorch, Middle-Rhine Valley, Germany). The drawing shows sedimentary and deformation structures, only locally exemplified, and mainly serves to clarify the relation between structures and their positions within the fold. In detail are shown: cross bedding in dm-thick quartzite layers indicating stratigraphic top, slump folding in these layers, variation of cross bedding geometry from fold limb to crest, thickness variation of bedding, intersection of second foliation and bedding, foliation fan in the fold crest, fold crest geometry, and variation of slickensides at different locations around the fold. The numbers refer to detail drawings of specific structures. In the drawing, fold axis and bedding of the long fold limb appear steep, in contrast to their true moderate dip of 31 to 35°. For better visibility of the lower surface of the bedding and its structures, the block is tilted in the drawing so that the direction of view is not horizontal but, rather, "obliquely upturned." Outcrop KR4142 (field book 35, Kruhl, 1995); size of original drawing = double page of the field book, that is, roughly A4; ballpoint pen; original German labeling replaced by English.

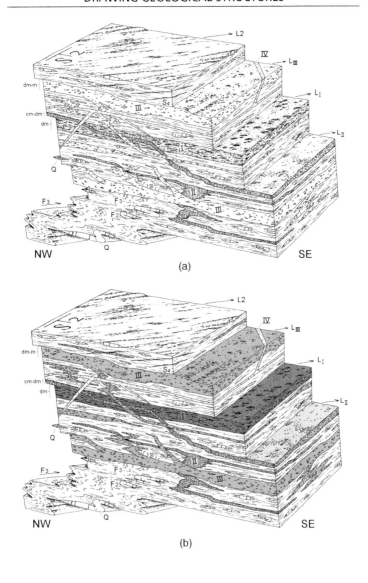

(a)

(b)

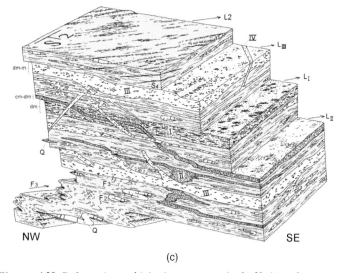

(c)

Figure 4.39 Deformation and injection structures in the Variscan basement of north-east Sardinia (Italy; Mte'e Senes); merged field drawings; reworked, digitally labeled, and filled; modified after Kruhl and Vernon (2005); size of original drawing ca. A4. (a) The plates of the echelon block are displaced in such a way that the foliation planes in the different granitoids I to III and in the metamorphic wall rocks, together with the orientations of the different lineations, are visible. Folded layers are attached to the block laterally and downwards so that the orientations of the fold axes, F3, can be recognized. Quartz-vein rods are slightly lifted out of the upper surface of the block, in order to illustrate their extension. The 3D orientation of late S4 foliation planes can be deduced from their intersection with two orthogonal vertical block faces. In the granitoids, only biotite platelets or aggregates of recrystallized grains of biotite in granitoid I are represented by short strokes, in order to avoid overloading the drawing and to keep it clear. Quartz vein lenses (Q) are characterized by cross fractures. The approximate thicknesses of granitoid I to III layers are given, because these thicknesses vary and cannot be inferred from the non-scaled drawing. (b) The foliated wall rocks are kept in light gray, leaving the internal fabrics visible and allowing a clear distinction from the granitoid layers. Thus, the granitoids can be more easily differentiated from one another and the chronology of injections is clearer. (c) A similar effect can be achieved by filling the four granitoid layers with different shades of gray.

Although it is generally not necessary to color individual areas of the drawing or fill them with different shades of gray, it can contribute to better visual separation of different layers and structures, and, thus, to an overall better readability of the drawing (Figure 4.39b, c). Here too, of course, care should be taken to ensure that the colors or shades of gray come close to the natural coloring of the rocks, or at least their relative lightness or darkness.

4.5 Rules of Labeling

The geological drawing represents a rich reservoir of detailed information about rocks and their structures from which further information about the condition and development of rocks and structures can be derived. Therefore, labeling should only be a supplement that is particularly appropriate where the drawing is not (cannot be) unique or information cannot be graphically conveyed. Too detailed labeling would counter the intention of the drawing, which is to provide information in a concise, easy, and quickly comprehensible manner.

This also means that the labels often consist of abbreviations—for rocks, minerals, and structures. The abbreviations are defined by the standard, and thus generally understandable, rules as found in textbooks or special literature—like Kretz (1983) and Whitney and Evans (2010) for minerals, for example. Since field drawings are usually aimed at storing data for personal use, there is nothing wrong with using simple, insightful abbreviations and developing one's own abbreviation language.

Measurements of planes and linear orientations are a central component of fieldwork. These measurements should be written directly on the structures in the drawing (Figures 4.8d, 4.11, 4.22, 4.24, 4.25, 4.34, 4.36, and 4.37). This is only possible if drawings contain free areas or the label is backed with white. In highly filled drawings or when detailed labeling is required, it is better to place the labels, at least partially, on the sides of the drawing (Figures 4.9d, 4.28, 4.33, 4.34, and 4.39). When starting a drawing, enough space should be reserved for this purpose as well. The closer the label is to the corresponding structure, the better one can link the qualitative and quantitative information when looking at the drawing.

If a structure has an unchanging orientation, it is enough to indicate one measured value. The stronger the variation, the more measured values, or at least two that indicate the range, should be specified (Figure 4.34 and 4.36). How exact a measured value is and how safely a structure or a sequence of structures can be classified, can only be assessed in the field, but not later "on the drawing board." Therefore, it is useful to mark these assessments in the drawing with the appropriate symbols, with "?" or "??" for an uncertain assessment and with "~" for an non-exact measurement, for example (Figures 4.19, 4.21, 4.22, 4.36, and 4.37).

Since the drawing of structures in the field is rarely (can rarely be) done to scale, a uniform scale can only be given for the fewest drawings, like when a single sample is drawn, for example (Figure 4.28). In some cases, the scale is not important for the geological message. In many cases, however, it is useful for the drawing to have one or (better) more scales, if the size of the structures is not clear from the form or other information, for example. For large overview drawings, like quarry walls or long road cuts, it is useful to specify the total size of the depicted outcrop. From there, individual structures can be fitted with scales to illustrate, for example, the thickness of layers or to clarify the size of individual geological bodies (Figures 4.20, 4.21, 4.26, and 4.37). It is often easier to integrate the scale into the label, like the size, or range in size, of crystals or the spacing between layers and foliation planes in millimeters, centimeters, or decimeters, for example (Figures 4.18, 4.19, 4.21, 4.23–4.25d, 4.33, 4.37, and 4.39). With this type of flexible scales, even unscaled drawings can contain precise information about the sizes of structures.

Even if the orientation of structures, and, thus, the orientation of the drawing, can be approximated from the measured values, it is easier for the viewer if the drawing is outfitted with cardinal points. During a review of the field drawings, that is, a screening of field data, this also allows for a quick and easy assignment of structures to large-scale, oriented processes (nappe transport, flow in magmatic bodies, etc.) and an understanding of the relation of the drawing to different outcrops. It is sufficient to indicate the directions in 45° increments, that is, as N, NE, E, and so on. In the form of numbers and with more precise degree-increments, the information would be too confusing. It is important that the cardinal points are set at the two ends of one side of the block-drawing and approximately horizontal, so that the orientation of the block can be clearly derived (Figures 4.18 to 4.26, 4.31 to 4.39).

A field drawing can help specify the exact location—which rock, which structure, which outcrop surface—where samples and photos were taken (Figure 4.18c). This is an important advantage. The exact assignment of samples to structures increases their value significantly and can even "save" a sample if it was mistakenly mislabeled or the label was partially or completely destroyed while sawing. The position of detail photos should, if possible, always be marked in the field drawing (Figure 4.21). Only then are such photos a good supplement to the drawing. The same is true for enlarged sections that accompany an overview drawing (Figures 4.27 and 4.37). Their position in the overview drawing should be precisely indicated so that the details can be clearly paired with the appropriate larger-scale structures.

Although it should go without saying, this is the moment to point it out again: Samples should always be taken with respect to their orientation. This means, the orientation of a sample surface should be measured and written on the drawing along with the strike- and dip-value and the sample number. Taking oriented samples requires only a little more effort, but yields many more benefits (Prior *et al.*, 1987).

In all that was discussed up until this point regarding labeling, the fact that the drawing must of course also include additional information about the outcrop, remained unmentioned. This includes a description of the location (with coordinates, if necessary), data for the map sheet (Figure 4.33), and possibly also information about the size, type, and condition of the outcrop (artificial/natural; fresh/weathered). In a series of outcrops, this information must not be specified for every outcrop, but it should be included somewhere. Vegetation and anthropogenic objects, on the other hand, do not belong in the drawing unless they are directly related to the geological message of the drawing.

No drawing is conceivable without outcrop and sample numbering. Of all the different ways to systematize this numbering, the simplest and shortest numbering is the best, in my opinion: The outcrop numbers run from 1 to infinity. "How much outcrop" is assigned one number, is one's own choice. For large, branched quarries or road cuts that are multiple hundred meters long, it may make sense to assign individual numbers for individual parts or sections. Every sample taken from an outcrop with a number is assigned exactly this number. If multiple samples are taken from one outcrop, a letter is added to the number. In this system, outcrops, samples and field drawing(s) are clearly linked to one another and with all data sets derived from a sample. With one number that, realistically, contains a maximum of four digits and one letter, all sorts of information can be bundled and retrieved quickly and accurately—a simple, clear, and effective system.

Summary

- When drawing in the field, speed, simplicity, and conciseness are paramount. We express our observations on paper in a concise language of symbols and make our field book into an information-rich and easy-to-read data archive.
- Most rock structures are spatially oriented and should therefore be drawn in 3D in a field book that is at least 20 x 15 cm (A5) and has, at most, an A4-sized drawing surface.
- Observation comes before drawing but does not stop when drawing has begun; the two processes constantly interact with one another.
- Structures in a field drawing are not represented the way they appear but, rather, straightened and grouped so that the drawing is uncluttered and the information can be clearly conveyed.

- A drawing is developed stepwise from large, simple structures to smaller, more complex ones.
- In order to make orientations of structures and fabrics on surfaces more visible, existing structures are drawn separately from the block or new structures are added.
- Enough space for "attachments" should be left around the perimeter of the drawing.
- The drawing should be loosely filled with internal fabrics. Leave gaps!
- The geological drawing is generally not made to scale. The size of individual structures is named or labeled with separate scales.
- Enlargements are good opportunities to combine structures of different scales.
- Labels consist of as many abbreviations as possible. In general, the drawing provides more information than what is understood by the label. The labels supplement and clarify the drawing.
- Measured values should be in the drawing and written close to the associated structures. Sampling and photo localities should also be marked in the drawing.
- Completely digital drawing is not a good idea. However, drawings can easily be improved, corrected, and cleanly labeled digitally.
- Showing an exact copy of reality is generally not the intention of a drawing. Instead, it should show a model of reality that clearly and effectively represents all relevant geological information.

■ **Exercise 4.1:**

Exercise 4.1 *The photo shows an orthogneiss from the Monte Rose Nappe (Beura, Val d'Ossola, Italy). The fine-grained biotite-quartz-feldspar ground-mass is weakly foliated and folded. The twofold deformation has also led to the cm-large plagioclase being plastically deformed and pulled apart in rods. Make a schematic block drawing of the sample. Take into account that the biotite layers consist of recrystallized grains in the sub-millimeter range and that the feldspar rods are covered with up to 1 mm wide white mica platelets. Complete the drawing in three steps. Draw only the outlines of the sample and the coarse structure in the first step. Second, make the fabric more precise. In the final step, fill and label the drawing.*

References

Coe, A.L. (ed.) (2013). Geological Field Techniques, Wiley-Blackwell, 323 pp.

Gill, R. (2010). Igneous Rocks and Processes, Wiley-Blackwell, 432 pp.

Gehlen, K. von and Voll, G. (1961). Röntgenographische Gefügeanalyse mit dem Texturgoniometer am Beispiel von Quarziten aus kaledonischen Überschiebungszonen. *Geologische Rundschau, 51,* 440–450.

Gosen, W. von (1982). Geologie und Tektonik am Nordostrand der Gurktaler Decke (Steiermark / Kärnten – Österreich). *Mitteilungen aus dem Geologisch-Paläontologischen Institut der Universität Hamburg 53,* 33–149.

Gwinner, M.P. (1971). Geologie der Alpen – Stratigraphie, Paläogeographie, Tektonik, E.Schweizerbart'sche Verlagsbuchhandlung, 477 pp.

Kretz, R. (1983). Symbols for rock-forming minerals. *American Mineralogist, 68,* 277–279.

Kruhl, J.H. (1984a). Metamorphism and deformation at the northwest margin of the Ivrea Zone, Val Loana (N. Italy). *Schweizerische Mineralogische und Petrographische Mitteilungen, 64,* 151–167.

Kruhl, J.H. and Vernon, R.S. (2005). Syndeformational emplacement of a tonalitic sheet complex in a late-Variscan thrust regime: fabrics and mechanism of intrusion, Monte'e Senes, northeastern Sardinia. *The Canadian Mineralogist, 43/1,* 387–407.

Nabholz, W.K. and Voll, G. (1963). Bau und Bewegung im gotthardmassivischen Mesozoikum bei Ilanz (Graubünden). *Eclogae Geologicae Helvetiae, 56/2,* 756–808.

Paterson, S.R., Vernon, R.H. and Tobisch, O.T. (1989). A review of criteria for the identification of magmatic and tectonic foliations in granitoids. *Journal of Structural Geology, 11,* 349–363.

Prior, D.J., Knipe, R.J., Bates, M.P., Grant, N.T., Law, R.D., Lloyd, G.E., Welbon, A., Agar, S.M., Brodie, K.H., Maddock, R.H., Rutter, E.H., White, S.H., Bell, T.H., Ferguson, C.C. and Wheeler, J. (1987). Orientation of specimens: essential data for all fields of geology. *Geology, 15,* 829–831.

Prothero, D.R. and Schwab, F. (1996). Sedimentary Geology, Freeman & Co., 575 pp.

Steck, A. (1968). Die alpidischen Strukturen in den Zentralen Aaregraniten des westlichen Aarmassivs. *Eclogae Geologicae Helvetiae, 61/1,* 19–48.

Steck, A. and Tièche, J.-C. (1976). Carte géologique de l'antiforme péridotitique de Finero avec des observations sur les phases de déformation et de recristallisation. *Bulletin suisse de Minéralogie et Pétrographie, 56,* 501–512.

Trewin, N.H. (ed.) (2002). The Geology of Scotland, *4th edition.* The Geological Society London, 576 pp.

Vernon, R.H. (2000). Review of microstructural evidence of magmatic and solid-state flow. *Electronic Geosciences, 5,* 2.

Vogler, W.S. (1987). Fabric development in a fragment of Tethyan oceanic litho-sphere from the Piemonte ophiolite nappe of the Western Alps, Valtournanche, Italy. *Journal of Structural Geology*, 9, 935–953.

Voll, G. (1960). New work on petrofabrics. *Liverpool and Manchester Geological Journal*, 2, 503–567.

Voll,G. (1976a). Recrystallization of quartz, biotite and feldspars from Erstfeld to the Leventina Nappe, Swiss Alps, and its geological significance. *Schweiz-erische mineralogische und petrographische Mitteilungen*, 56, 641–647.

Whitney, D.L. and Evans, B.W. (2010). Abbreviations for names of rock-forming minerals. *American Mineralogist*, 95, 185–187.

Zurru, M. and Kruhl, J.H. (2000). Die Loreley. Steinalt und faltig – jung und schön! Selden & Tamm, 70 pp.

5

GEOLOGICAL STEREOGRAMS

5.1 Foundations

Structures in larger regions can be grouped into geological stereograms. These are created in much the same way as 3D drawings of specimen sections and outcrops, but summarize, usually along profiles, the distribution of structures in larger regions. Generally, the same rules as for 3D drawings apply: (i) Structures are only represented schematically and symbolically, and at different scales. (ii) Block surfaces are oriented parallel to the important planar or linear structures. (iii) Geological stereograms are designed like echelon blocks, in which certain structures are exemplarily drawn, that is, continued, from the inside to the outside of the block to make them more visible. (iv) The block is oriented so that as many surfaces as possible are clearly visible. (v) The surfaces are only loosely and fractally filled with internal structures. (vi) Even in a geological stereogram, individual parts may be separated from others for better representation.

The simplified representation of larger structures by which principles of large-scale geological structures, often a mountain range or section of a range, can be clarified, are one advantage of geological stereograms. On the other hand, the variation of structures and rocks over large distances and in large areas can be represented quite well in detail. Two-dimensional cross-sections through mountains belong to the classic inventory of geological representation and date back to the early days of geological literature (Geikie, 1893, Peach & Horne, 1914, Heim, 1919-22, Staub, 1924). Even the third dimension was often incorporated into representations—(i) in the form of parallel profiles (Heim, 1921), (ii) as orthogonal or transversal profiles (Matter *et al.*, 1980, figure V.8), (iii) as multiple, spaced surfaces or layers, specifically suitable for large-scale complex folding (Ramsey, 1967, chapter 10, Passchier *et al.*, 1981, Holdsworth & Roberts, 1984), or (iv) as "real" geological stereograms (Hagen, 1969). If the block surface schematically models the morphology, the influence of geology on the surface forms can be seen (Trümpy, 1980, Wagenbreth & Steiner, 1990). Some parts of mountain ranges are not only presented in slender stereograms, but in multiple adjacent echelon blocks with depthwise representation and partially covered structures (De Römer, 1961, Stephenson & Gould, 1995, figure 19).

Drawing Geological Structures, First Edition. Jörn H. Kruhl.
© 2017 John Wiley & Sons Ltd. Published 2017 by John Wiley & Sons Ltd.

Since all of these stereograms are scaled, they are missing, as opposed to those shown in this book, the medium and smaller structures. Moreover, despite the cross sections and partially lifted layers, they remain relatively "closed" blocks. In addition, the advantages of drawing structures exemplarily, that is, continuing them from the inside to the outside of the block, are not used. The first unscaled geological stereograms that yielded a much higher information content through the enlarged representation of small-scale structures, the shifting or lifting of individual block parts relative to one another and continuing structures towards the outside, were introduced by Gerhard Voll in the early 60s (Voll, 1960, Voll, 1963, Nabholz & Voll, 1963) and taken up by other authors, albeit sporadically, in later years (Kruhl, 1979, 1984a,1984b, Vogler & Voll, 1981, v. Gosen, 1982, 1992, Vogler, 1984, Altenberger *et al.*, 1987, v. Gosen *et al.*, 1990). By combining large-scale with unscaled, small-scale structures, they provide a significantly deeper insight into the make up and structural development of larger regions. Even if the geological stereograms found in literature extend almost exclusively horizontally, this need not necessarily be the case. A vertical orientation can, for example, represent the local composition and structural development of the entire continental crust in a "handy" manner (Voll, 1983).

A major difference between geological stereograms and 3D sample or outcrop drawings is that stereograms are entirely constructed, from the outset, with help of existing data (mainly field drawings). The necessary space can therefore be estimated *a priori* without having to add structures retrospectively. One can decide which structures to remove and can immediately orient the diagram so that all surfaces are optimally visible. The of field drawings typical "forcible" addition of structures that were recognized (too) late is thus avoided. Some geological stereograms that appear in literature were originally created in an up to A0 big area and are greatly downsized for publication purposes (Voll, 1963, figure 5, Nabholz & Voll 1963, figure 10 and 11, Voll, 1976b, plate 1). This is how drawings with the finest line weights in which lots of details are housed are created.

However, these extensive amounts of details can also be a disadvantage. The desire for the most encompassing representation of all structures conflicts with the viewer's desire for clarity and simplicity. As beautiful and aesthetically complex representations sometimes are, simplicity and the resulting intelligibility are preferred, when in doubt.

Although a geological stereogram encompasses a larger area than a 3D sample or outcrop drawing, the structures need not be reduced further or schematized. A geological stereogram is also created at different scales, albeit spanning a correspondingly larger scale range, and is a simplification of outcrop sketches. It should depict the main structures, but those must not necessarily be major regional structures. What matters is not the size of a structure but, rather, its meaning.

5.2   Orientation and Extension

Geological stereograms almost always represent geological areas with a prominent longitudinal extension. This is the case in all zones created through the collision of plates or continental crusts: mountains, subduction zones, exhumed regions, fault zones, volcanic belts, and so on. The geological stereogram extends across the strike of the layered structures in the zone across distances of a few hundred meters to several kilometers. Height and width of the block depend on how many of the structures' details are to be represented. Usually, the height of the stereogram roughly corresponds to its width. In the simplest case, plates, each of which represent a rock unit, are strung together and shifted to one another as in an echelon block. For better visibility, structures are drawn out separately below. In a way, a geological stereogram is not much more than an extended echelon block (Figure 5.1).

In contrast to a sample or outcrop drawing, the much larger stereogram can be linked to the large-scale morphology of a region. Therefore, in a first step, a geological profile with true length, thickness, and surface morphology can be constructed, showing the realistic thicknesses and shapes of rock layers or rock units. In a second step, this profile is extended to 3D and to an echelon block with the characteristics of an unscaled stereogram, like enlarged small-scale structures or the continuations of planar structures towards the outside (Figure 5.2). Thus, a closer link to the "real" geological situation is preserved.

If the zone being represented includes many outcrop gaps, they must not all appear in the geological stereogram. If only the representation of different structures and rocks is important and their sequence is irrelevant for the interpretation, the structures could be strung together without gaps. This also increases the clarity of the representation. The specification of places or prominent localities helps to geographically classify individual parts of the geological stereogram (Figure 5.3).

Now and then, different structures, that are to be shown extensively, are developed on the flat faces of the individual plates. Larger areas of these faces can be made visible by offsetting individual plates not vertically, but instead horizontally, from one another (Figure 5.4). A side effect of this representation is that the geological stereogram stretches not transversely, but, rather, parallel in space-saving fashion along the long side of the drawing area with equally good visibility of all three orthogonal block surfaces.

If structures that extend not only horizontally, but also depthwise, are to be represented in a geological stereogram, the form of the diagram must be adapted accordingly. A rather irregularly shaped block, adapted in its contours to the structures that are to be represented, results. The usual options of the echelon block drawing—mutually shifted plates, structures continued from the inside to the outside of the plate laterally and below the drawing—are complemented by a

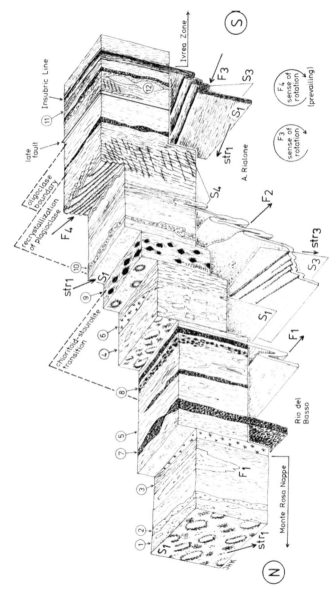

Figure 5.1 North-south–oriented geological stereogram of the Sesia Zone between Ivrea Zone and Monte Rosa Nappe, west of Val Loana (Western Alps, Italy). It covers a total length of ca. 5 km and represents the principal rock types and structures of the zone schematically. The stereogram is constructed as an elongate echelon block. Each plate of the block comprises one to several rock units. The plates are inclined and displaced against each other in such a way that the three orthogonal faces are equally well visible. Folded layers are extended downwards to visualize the orientations of the fold axes. In order to avoid overloading of the stereogram by labeling, numbers relate to the various rock types described in an additional text. This text also includes the orientations of the deformation structures. The planes of the beginning of plagioclase and K-feldspar recrystallization, and of the chloritoid-staurolite transition represent the spatial distribution of the maximum temperature of metamorphism. In addition, the general sense of folding of the third and fourth deformation event is given, because it cannot be readily inferred from the drawing. The localities of one small river and one alp are indicated for better geographic positioning of the stereogram. Size of original drawing ca. A3.

Figure 5.2 North-northwest–south-southeast–oriented geological stereogram through the northern margin of the fossil Variscan lower continental crust of Calabria (Italy) and the northerly adjacent Castagna Unit. The construction of the stereogram started with the profile A-B showing the surface morphology, and the realistic thickness and orientation of rock units 1 to 6 in depth, based on surface geology. The rock structures are schematized and not to scale. Their different densities lead to a variably dark appearance of the rocks, roughly analogous to their relative light-dark contrast in nature. In a second step, the profile is extended to a 3D echelon block that allows the orientations of planar structures (layering, foliation planes) and folds in 3D to be represented. Most importantly, the variable orientations of lineations of the different deformation events can be shown on the surfaces of single plates, which are parallel to different schistosities. In order to represent small-scale structures more clearly, schematized sample and outcrop drawings are placed below the stereogram. These drawings show structures that cannot be represented in the stereogram due to the limited thickness of the rock layers. Labeling is kept to a minimum to maintain the clarity of the stereogram, and all necessary additional information is given in an accompanying text. This geological stereogram represents the complex deformation history of the region, as well as the numerous and differently scaled rock structures, in a relatively simple and instructive way. The location of the profile A-B is marked in a geological map. Size of original drawing ca. A3; black ink pen with line weight of 0.25 mm; based on unpublished data (Kruhl, 1996).

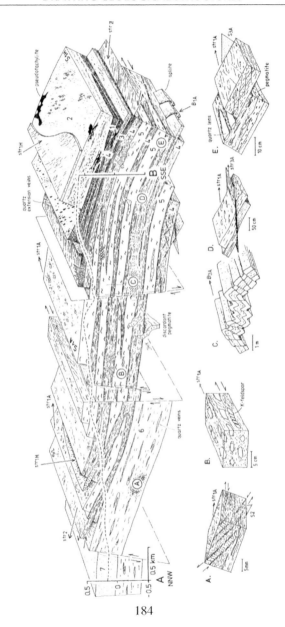

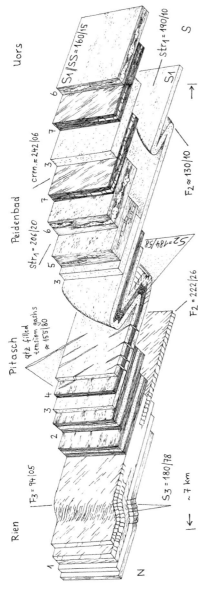

Figure 5.3 *North–south–oriented geological stereogram from the Permian-Trias cover of the eastern Gotthard Massif (Valser Rhine valley, Switzerland), partially constructed on the basis of Figures 4.33 to 4.36. Despite larger gaps of exposure, the stereogram is continuous in order to (i) condense the drawing, (ii) illustrate the general orientation of layers, and (iii) show the consistent style of structures. The labeling is reduced to a minimum. The different rock types are marked by numbers, which are explained in an accompanying text. In general, the stereogram needs detailed explanations. The localities of four small towns are given for better geographic positioning of the stereogram. Thereby, the partially systematic changes of fabrics (e.g., the first stretching lineation, str1) and of the lithology in S-N extension can be spatialized more effectively. The total length of the stereogram is 7 km; size of the original drawing ca. A3.*

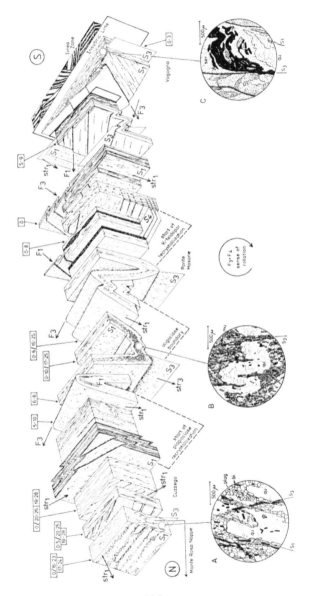

backward extension of the diagram by which other, differently oriented structures can be visualized (Figure 5.5). When dealing with rounded bodies, like a pluton with its contact aureole, for example, individual parts can be cut out to allow a glimpse of individual layers and an in-depth look at differently oriented structures (Figure 5.6). Such representations have nothing to do with the morphology and appearance of the bodies in the field. They inform, in a concise manner, about the nature, spatial occurrence, and sequences of the structures.

Even in large, but compact, geological stereograms, complex fold structures and fold axes that change their orientations, can be represented only inadequately. If an unobscured view from different directions and on different areas is desired, the entire diagram must be kept as "loose" as possible. This is best achieved if only a few thin layers are drawn and, if necessary, separated by open space. Small spatial variations in the folds can be represented in this way. In this type of representation, the drawing clearly grows depthwise or stretches vertically. Through loosely stacked layers and folds, many visible surfaces emerge and ways of representing different structures, like axial folds, in their spatial variation, for example, are

◄──────────────────────────────────────

Figure 5.4 *Approximately north-south–oriented geological stereogram of the Sesia Zone in Val d'Ossola between Ivrea Zone and Monte Rosa Nappe (Western Alps, Italy). It covers a total length of ca. 7 km and schematically illustrates the main rocks and structures of the Zone. The stereogram is constructed as an extended echelon block with each plate representing one rock unit. The plates are vertically and laterally displaced against each other so that the three orthogonal surfaces are similarly facing the viewer. Mostly the flat surfaces are well visible. Consequently, lineations and fold crests are clearly viewable and foliation planes and folded layers do not need to be continued from the inside to the outside of the plates. The calcium content ("anorthite content") of host and recrystallized grains of plagioclase is given and related to specific localities. The planes of the beginning of plagioclase and K-feldspar recrystallization, and the oligoclase boundary indicate the northward increasing maximum temperature of metamorphism and its spatial distribution. The stereogram clarifies that these planes dip more flatly to the north than the layering. In order to avoid excessive labeling of the stereogram, the orientations of deformation structures are shown in an additional text, together with the names of the different rock types. The general sense of folding of the third and fourth deformation event is given in addition, because it cannot be readily inferred from the drawing. Thin-section drawings of three typical rocks complement the larger fabrics represented in the stereogram. Thus, the stereogram covers a scale ranging from micrometer to kilometer. Three localities are given for better geographic positioning of the stereogram. Size of original drawing ca. A2; modified after Altenberger* et al. *(1989).*

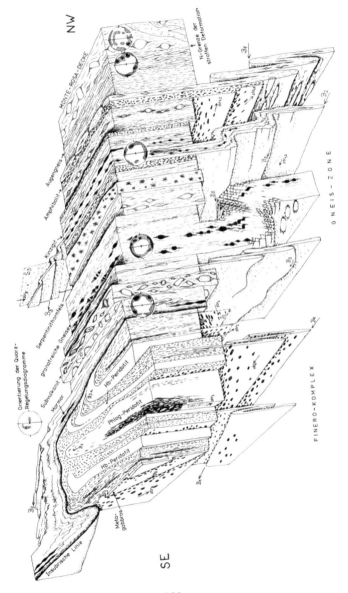

188

Figure 5.5 Schematic, *northwest-southeast–oriented geological stereogram through the ultrabasic, magmatic Finero Complex and the northward bordering Sesia Zone (Western Alps, Italy). Folded layers are drawn downward and sideward out of the block to show the orientations of fold axes and lineations. The southeastern part of the stereogram is extended to the southwest, in order to illustrate the large fold at the end of the Finero Complex. To increase the contrast within the Finero Complex, only gabbro, but no peridotite, layers are filled with internal fabrics, except for a single ultramylonite zone (black) and the schistosity (short strokes) in a fold bend. Internal fabrics are only drawn as far as to emphasize the characteristics of the different rocks. In comparison to the thickness of the layers, crystals are extremely enlarged to show their typical shapes. Diagrams of quartz c-axis orientations are given for some of the rocks. The orientations of the different structures are not affixed to the stereogram but included in an accompanying text. Total length of the stereogram in SE-NW extension ca. 4 km; size of original drawing ca. B3; from Kruhl (1979).*

created (Figure 5.7). The danger of such a complex geological stereogram is obvious: The drawing can become overloaded with too many details, and it is burdensome for the viewer to recognize the structural diversity, perceive the relationships, and generally retain an overview of "what is going on geologically."

5.3   Additions and Labeling

Much more so than in sample and outcrop drawings, the geological stereogram must be outfitted with information beyond the drawing. The stronger the graphical simplification, the harder it is to definitively recognize structures and rocks by their shape and internal fabric at larger scales, and the more important detailed and clear labels become. However, this can quickly lead to a drawing that is overburdened, with rock names, for example. This can most easily be avoided by numbering and shifting the labels to the figure caption (Figure 5.1). Even then, the essential structures should remain labeled to retain good readability of the drawing. Should the stereogram cover multiple geological units, the boundaries of these units should of course be marked as separate areas in the stereogram (Figures 5.1 and 5.4).

A geological stereogram spans over a significantly greater distance than a sample or outcrop drawing. This also means that changes in rock properties that don't appear in small areas are visible—like the degree of metamorphism in metamorphic rocks, but also the changes of facies in sedimentary rocks, or the chemistry and mineralogy in large intrusive bodies, for example. Often, this information cannot be relayed graphically; instead, a predominantly numerical representation is

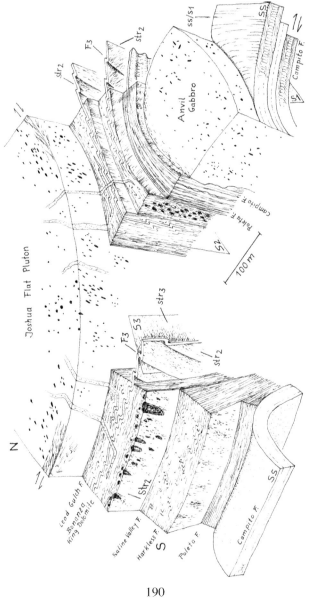

←

Figure 5.6 Schematic geological stereogram of the Joshua Flat Pluton, the Anvil Gabbro, and the intercalated metasedimentary rocks (Inyo Mountains, California). The stereogram serves to illustrate the intense deformation of the metasedimentary rocks during the emplacement of the two plutons. The roundish shape of the plutons and their relative position require a "branched" shape of the stereogram. This open configuration visualizes the three orthogonal principal structural planes equally well. It also facilitates the drawing of folded layers out of the block and illustrates the orientations of fold axes. The stepped arrangement of the different rock layers creates additional surfaces parallel to the main schistosity, which allow particularly the orientations of various lineations to be represented. The rare structures in the two plutons are only drawn as far as necessary. In contrast, the metasedimentary rocks are completely filled with internal fabrics, in order to adequately represent the numerous variable structures. Only those structures are labeled, whose denotation would otherwise be ambiguous. The 100 m scale applies only to the thickness of the metasedimentary rocks. The internal fabrics of these rocks are drastically scaled up and only represented exemplarily. Names of rock units are attached to the different plates of the stereogram. This builds a bridge between the schematic, simplified stereogram and the comparatively precise geological map of the region. Compiled from field drawings (field books 33 and 34, Kruhl, 1991, 1992); size of the original drawing A3.

required. The change in degree of metamorphism, for example, can be reflected in the variation of calcium content ("anorthite content") in plagioclase (Figure 5.4).

The occurrence of mineral reactions, or the boundaries between the distribution of different mineral parageneses, can, however, be integrated into the geological stereogram in the form of lines or surfaces (Figures 5.1 and 5.4). If the 3D orientations of the boundaries are known, they can deliver, in combination with the orientation of rock layers or structures, valuable information about the deformation and metamorphic history of the regions pictured in the geological stereogram. In basement regions with different deformation phases, it can be useful to specify the sense of rotation of different folding events (Figures 5.1 and 5.4).

Even diagrams can be used to increase the information content of a geological stereogram. If the crystallographic arrangement of minerals is being measured, it makes sense to assign the diagrams to localities of measured samples and to group them accordingly in the stereogram (Figure 5.8). If space allows, the diagrams can be inserted into the drawing, if possible, so that their orientation matches the orientation of the structures (Figure 5.5). If the orientation of the diagrams is also based on the orientation of structures, the (experienced) viewer can directly derive further valuable information (about shear sense, for example).

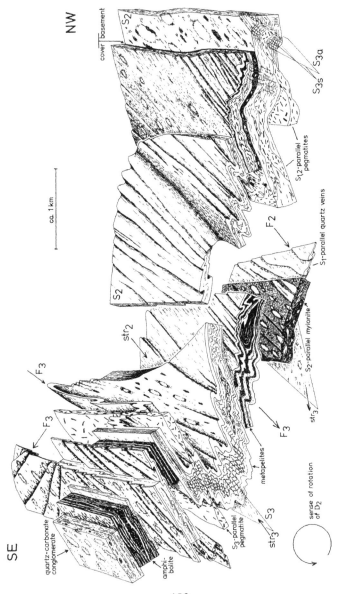

Figure 5.7 Schematic, northeast-southwest–oriented, ca 7 km long geological stereogram of the Precambrian basement, overlain by the Helgeland Complex, northwest of Grong, between Terråk und Leka (65th latitude circle, Northern Norway). The principal structures comprise km to 100 m large folds with, in part, significantly different axial orientations. Only single rock layers are drawn and partly lifted against each other, in order to illustrate the shapes of the folds and the orientations of the axes as well as the different lineations. Lineations also serve to emphasize the 3D shape of the folds. In addition, the general sense of folding of the second deformation event is given, because it cannot be readily inferred from the drawing. The orientations of the different structures are included in an accompanying text. Size of the original drawing ca. A2; modified after Kruhl (1984b).

For exact representation of small structures, it makes sense to integrate thin section drawings into the geological stereogram (Figure 5.4). This is worthwhile if the drawings deliver important information that could otherwise not be depicted graphically due to the scale of the stereogram. There are, in principle, no limits on the integration of additional information. Graphic information, especially, should be inserted directly into the geological stereogram if it provides evidence of large-scale 3D variation of fabrics or other rock properties.

5.4   Digital Processing

What applies to sample and field drawings is also true for the generation of geological stereograms. Specifically, presentations for publications that require cleanliness and clarity can benefit from digital processing. Although it is generally not necessary to color individual areas of the stereogram or fill them with different shades of gray, filling individual layers in such a way represents a simple and effective way to improve clarity and readability of stereograms that were, apart from that, produced manually. Thus, again, advantages of both methods can be combined. Good examples of manually generated and digitally processed geological stereograms can be found in the literature (v. Gosen, 2002, 2009).

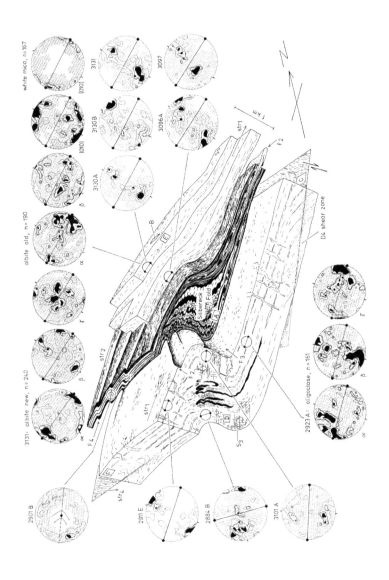

Figure 5.8 Schematic geological stereogram of the Silbereck sequence (with the large D3 Silbereck fold) and of the adjoining gneisses of the Variscan basement; northeast margin of the Tauern Window (Eastern Alps, Austria). In contrast to their appearance in the field, the metasedimentary rocks are kept dark, in order to highlight the thin layers within the gneisses and the shape of the fold. The structures of the deformation events D1 to D4 are sketched only locally. The orientations of biotite crystal aggregates (short strokes) indicate the orientations of the D1 and D4 lineations in the gneisses. In addition to the lithological and structural inventory, the preferred crystallographic orientations of quartz, plagioclase and white mica are represented and related to different tectonic locations around the Silbereck fold. The diagrams are positioned in such a way that they relate to the orientations of structures (foliations, lineations) in the rocks. The km-sized, monoclinic Silbereck fold, like some minor D2 folds, documents top-to-the-ENE transport during D3. Uppercase letters refer to locations of thin-section photographs. The stereogram is constructed based on numerous outcrop drawings and measurements. The km-scale applies to the thickness of the layers; size of original drawing ca. A3; modified after Kruhl (1993).

Summary

- Structures in larger areas can be combined to create geological stereograms according to the same rules that apply to 3D drawings in the field.
- Intentions of the geological stereogram: to show the geological structure of larger areas and clarify the variation of rocks and structures across large distances.
- When depicting complex, large and numerous, small structures, geological stereograms can be constructed as open and variably formed blocks.
- Like an outcrop drawing, the geological stereogram is also created at different scales. It is just that the scales span a correspondingly larger area.
- A geographical stereogram must be labeled more extensively than a field drawing.
- The graphic representation of microfabrics and data about other rock properties are incorporated directly into the geological stereogram as important additional information.
- Geological stereograms should be detailed but not too complex.

References

Altenberger, U., Hamm, N. and Kruhl, J.H. (1987). Movements and metamorphism north of the Insubric Line between Val Loana and Val d'Ossola, N.Italy. *Jahrbuch der Geologischen Bundesanstalt Wien*, 365–374.

De Römer, H.S. (1961). Structural elements in southeastern Quebec, northwestern Appalachians, Canada. *Geologische Rundschau*, 51, 268–280.

Geikie, A. (1893). Text-Book of Geology, 3rd edition, MacMillan & Co., 1147 pp.

Gosen, W. von (1982). Geologie und Tektonik am Nordostrand der Gurktaler Decke (Steiermark / Kärnten – Österreich). *Mitteilungen aus dem Geologisch-Paläontologischen Institut der Universität Hamburg*, 53, 33–149.

Gosen, W. von (1992). Structural Evolution of the Argentine Precordillera: the Rio San Juan section. *Journal of Structural Geology*, 14, 643–667.

Gosen, W. von, Buggisch, W. and Dimieri, L.V. (1990). Structural and metamorphic evolution of Sierras Australes, Buenos Aires Province, Argentina. *Geologische Rundschau*, 79, 797–821.

Hagen, T. (1969). Report on the geological Survey of Nepal, Volume 1 – Preliminary Reconnaissance, Denkschriften der Schweizerischen Naturforschenden Gesellschaft LXXXVI/1, reprinted by Nepal Geological Society, 2013, 199 pp.

Heim, A. (1919-22). Geologie der Schweiz, 2 Vol., Tauchnitz, Leipzig.

Heim, A. (1921). Geologie der Schweiz, Vol. II-1, Tauchnitz, Leipzig, 476 pp.

Holdsworth, R.E. and Roberts, A.M. (1984). Early curvilinear fold structures and strain in the Moine of the Glen Garry region, Inverness-shire. *Journal of the Geological Society London*, 141, 327–338.

Kruhl, J.H. (1979). Deformation und Metamorphose des südwestlichen Finero-Komplexes (Ivrea-Zone, Norditalien) und der nördlich angrenzenden Gneiszone, Doctoral Thesis, The Faculty of Mathematics and Natural Sciences, Rheinische Friedrich-Wilhelms-Universität Bonn, 142 pp.

Kruhl, J.H. (1984a). Metamorphism and deformation at the northwest margin of the Ivrea Zone, Val Loana (N. Italy). *Schweizerische mineralogische und petrographische Mitteilungen*, 64, 151–167.

Kruhl, J.H. (1984b). Deformation and metamorphism at the base of the Helgeland nappe complex, northwest of Grong (Northern Norway). *Geologische Rundschau*, 73,735–751.

Kruhl,J.H. (1993). The P-T-d development at the basement-cover boundary in the north-eastern Tauern Window (Eastern Alps): Alpine continental collision. *Journal of metamorphic Geology*, 11, 31–47.

Matter, A., Homewood, P., Caron, C., Rigassi, D., Stuijvenberg, J. van, Weidmann, M. and Winkler, W. (1980). Flysch and Molasse of Western and Central Switzerland, in: R. Trümpy (ed.), Geology of Switzerland – A Guide-Book, Part B, Geological Excursions, 261–293.

Nabholz, W.K. and Voll, G. (1963). Bau und Bewegung im gotthardmassivischen Mesozoikum bei Ilanz (Graubünden). *Eclogae Geologicae Helvetiae, 56/2,* 756–808.

Passchier, C.W., Urai, J.L., Van Loon, J. and Williams, P.F. (1981). Structural geology of the central Sesia Lanzo Zone, Geologie en Mijnbouw, 497–507.

Peach, B.N. and Horne, J. (1914). Guide to the Geological Model of the Assynt Mountains, Geological Survey and Museum, H.M.S.O., 32 pp.

Ramsey, J.G. (1967). Folding and Fracturing of Rocks, McGraw-Hill, 568 pp.

Staub, R. (1924). Der Bau der Alpen, Beiträge zur Geologischen Karte der Schweiz, N.F. 52, Bern.

Stephenson, D. and Gould, D. 1995. The Grampian Highlands, 4th edition, British Regional Geology, British Geological Survey, HMSO London, 261 pp.

Trümpy, R. (1980). Geology of Switzerland – A Guide-Book, Part A: An Outline of the Geology of Switzerland, Schweizerische Geologische Kommission, Wepf & Co. Publishers, 104 pp.

Vogler, W.S. (1984). Alpine structures and metamorphism at the Pillonet Klippe – a remnant of the Austroalpine nappe system in the Italian Western Alps. *Geologische Rundschau, 73,* 175–206.

Vogler, W.S. and Voll, G. (1981). Deformation and metamorphism at the south-margin of the Alps, east of Bellinzona, Switzerland. *Geologische Rundschau, 70,* 1232–1262.

Voll, G. (1960). New work on petrofabrics. *Liverpool and Manchester Geological Journal, 2,* 503–567.

Voll, G. (1963). Deckenbau und Fazies im Schottischen Dalradian. *Geologische Rundschau, 52/2,* 590–612.

Voll, G. (1976b). Structural studies of the Valser Rhine valley and Lukmanier region and their importance for the nappe structure of the Central Swiss Alps. *Schweizerische mineralogische und petrographische Mitteilungen, 56,* 619–626.

Voll, G. (1983). Crustal xenoliths and their evidence for crustal structure underneath the Eifel volcanic district, in: Plateau Uplift, K. Fuchs et al. (eds.), Springer.

Wagenbreth, O. and Steiner, W. (1990). Geologische Streifzüge, 4th edition, Deutscher Verlag für Grundstoffindustrie, 204 pp.

6

SOLUTIONS

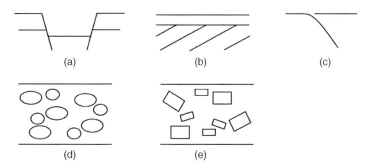

(a) (b) (c)

(d) (e)

Exercise 1.2 *(a) Graben: The upper part of a graben structure depicts its characteristics clearly enough. Seven short, straight strokes suffice for the pictogram. The fact that the two normal faults of the graben are usually listric and merge with horizontal shear planes at greater depths does not need to be shown and would overcrowd the sketch. The horizontal, double lines are important, because they illustrate the depression of the layers in the graben. **(b)** Unconformity: It is enough to represent a few dipping layers covered by only one horizontal layer. **(c)** Subduction: Representations of cross sections through subduction zones appear in the literature on numerous occasions and always according to the same schema. Therefore, two lines are enough to depict the subducting plate and its lid. There is hardly another pictogram that is so sparse yet still so expressive. **(d)** Conglomerate: In addition to the ellipses for the pebble representation, the layer needs to be bounded even if it's not typical for a conglomerate. Without a boundary, the sketch would not be clearly interpretable. The number of pebbles could be slightly lower. The fractal distribution of size, rather than the size itself, is important. It makes the sketch more natural. **(e)** Porphyritic Granite: The phenocrysts are represented by rectangles. Other than that, the same applies as to the conglomerate: A clear layer boundary is necessary.*

Drawing Geological Structures, First Edition. Jörn H. Kruhl.
© 2017 John Wiley & Sons Ltd. Published 2017 by John Wiley & Sons Ltd.

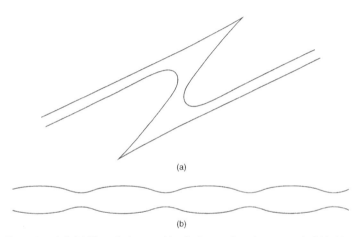

(a)

(b)

Exercise 1.6 (a) *The solution provides the impression of two strongly folded layers: The weak layer that lies between the other two has a point on the outside and a soft inner bend.* ***(b)*** *A slight contrast in strength leads to only a slight boudinage with appropriately softly curved layer boundaries.*

(a)

Exercise 2.1 (a) *To illustrate the folding, it suffices to mark the longitudinal surfaces of select biotite crystals with short lines. To depict the fold crests, the lines there are closer together. The quartz areas are kept blank. Ore exsolved from biotite is represented by somewhat thicker lines.*

(b)

Exercise 2.1 *(continued)* ***(b)*** *The higher line density in the biotite layers clarifies the kink and annealing of the biotite platelets and enhances the contrast to the quartz areas. It also better highlights the thin-layer structure of the rock.*

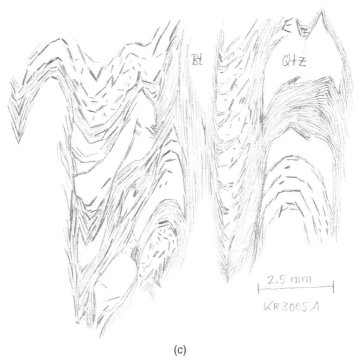

(c)

Exercise 2.1 *(continued) **(c)** Even a quick sketch should illustrate the contrast between biotite and quartz layers, and emphasize the kink of the biotite layers. However, in light of saving time, it is sufficient to represent the biotite with longer solid lines and leave the sketch incomplete.*

203

Exercise 2.2 The contrast between the four minerals can be established without a problem. The line weights of garnet (reinforced by peripheral dotting), pyroxene/amphibole, and plagioclase are different enough. Through dense internal dotting, which mimics its inherent color, amphibole further distinguishes itself from pyroxene. A few crystals, whose angles between cleavage planes are typical of amphibole and pyroxene, amplify the difference even more. The lamination with deformation twins, typical of plagioclase in metamorphic rocks, is another characterizing feature.

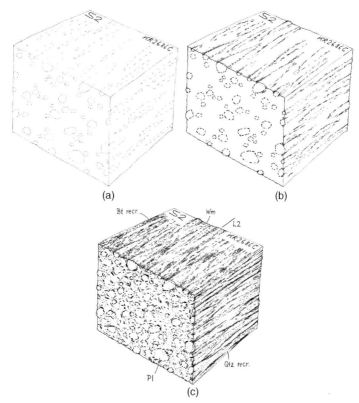

Exercise 4.1 (a) In the first step, the outline of the sample is sketched slightly perspectively as a rectangular block and the outlines of the feldspar rods are marked with dashed lines. For logical reasons, the surfaces are labeled now before they are filled with internal fabrics. (b) The slight lifting of select rods from the block and the shading with short lines help strengthen the 3D appearance of the drawing. The lines also represent the recrystallized biotite blobs. (c) In the last step, primarily the internal fabrics in the cross section are added to the rods. They also serve to highlight the weak, open folding of the equally weak first foliation. To avoid overfilling the drawing, the fine-grained groundmass is not represented.

INDEX

Note: numbers in *italics* refer to numbers of drawings; *E* refers to exercises.

Drawing Geological Structures, First Edition. Jörn H. Kruhl.
© 2017 John Wiley & Sons Ltd. Published 2017 by John Wiley & Sons Ltd.